GOVERNING MILITARY SACRIFICE

Governing Military Sacrifice

Drones, Privatization, and the Future of War

BIANCA BAGGIARINI

UNIVERSITY OF TORONTO PRESS
Toronto Buffalo London

Toronto Buffalo London
utppublishing.com
Printed in the USA

ISBN 978-1-4875-1088-6 (cloth)
ISBN 978-1-4875-1205-7 (EPUB)
ISBN 978-1-4875-1161-6 (PDF)

Library and Archives Canada Cataloguing in Publication

Title: Governing military sacrifice : drones, privatization, and the future of war / Bianca Baggiarini.
Names: Baggiarini, Bianca, author.
Description: Includes bibliographical references and index.
Identifiers: Canadiana (print) 2025030399X | Canadiana (ebook) 20250304058 | ISBN 9781487510886 (hardcover) | ISBN 9781487511616 (PDF) | ISBN 9781487512057 (EPUB)
Subjects: LCSH: Drone warfare. | LCSH: Sacrifice. | LCSH: Privatization. | LCSH: War – Forecasting.
Classification: LCC UG1242.D7 B34 2026 | DDC 623.74/69 – dc23

Cover design: John Beadle
Cover image: sibsky2016/Shutterstock.com; Nongkran/Adobe Stock

We wish to acknowledge the land on which the University of Toronto Press operates. This land is the traditional territory of the Wendat, the Anishnaabeg, the Haudenosaunee, the Métis, and the Mississaugas of the Credit First Nation.

This book has been published with the help of a grant from the Federation for the Humanities and Social Sciences, through the Awards to Scholarly Publications Program, using funds provided by the Social Sciences and Humanities Research Council of Canada.

University of Toronto Press acknowledges the financial support of the Government of Canada, the Canada Council for the Arts, and the Ontario Arts Council, an agency of the Government of Ontario, for its publishing activities.

Canada Council for the Arts
Conseil des Arts du Canada

Funded by the Government of Canada
Financé par le gouvernement du Canada

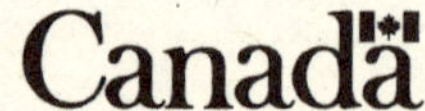

Contents

GOVERNING MILITARY SACRIFICE

Chapter 1

Why Sacrifice?

One must urge the citizen-soldier to give up his life so that a particular way of life may be reproduced – a sacrifice.

– Asad (2007, p. 85)

Does War Need Sacrifice?

On 29 January 2017, just five days after newly elected US president Donald Trump took office, a Special Forces raid in the middle of the night on a village in rural Yemen went terribly wrong. The goal of the covert counterterrorism operation was to acquire intelligence. Yet chaos ensued as people attempted to flee the gun battle before helicopters opened fire and "shot at everything," including homes and those fleeing (Al Sane & Shabibi, 2017). The raid came on the heels of weeks of surveillance, Navy SEALs rehearsals in nearby Djibouti, and the agreement that such a mission would have to be precise. This was due to the makeup of the targeted Yemeni neighbourhood, which was heavily guarded, and was made up of both civilian homes and militant bases. Lacking sufficient intelligence, ground support, and adequate backup preparations, interview data with unnamed military officials reveals that the trouble for the US forces started when their cover was blown. One aircraft crash-landed, injuring three. Moreover, "one thing after another went wrong from the start of the mission … the Special Forces were confronted by heavily fortified positions, including landmines, and faced heavy gunfire from buildings all around during the fifty-minute firefight" (Ackerman, Burke, & MacAskill, 2017). An eleven-year-old, who, upon being awoken, went outside to investigate, was the first to be killed. Despite the killing of twenty-five civilians, including a three-month-old baby who was asleep in her crib (totalling nine children under the age of thirteen) and the destruction of a US$70 million aircraft, Trump's press secretary Sean Spicer called it a successful operation, claiming it "achieved the purpose it was going to get – save the loss of life that we suffered and the injuries that occurred. The goal of the raid was intelligence-gathering, and that's what we received, and that's what we got" (Al Sane & Shabibi, 2017).

According to the previously mentioned reports, it was the presence of drones that initially tipped the Yemeni villagers off. Drones are now a central capability of any modern or modernizing military (Medina, 2014; Sauer & Schörnig, 2012). In essence, a drone is an artefact that *extends* the sensibilities and violent capabilities of a human soldier, while affording the opportunity to physically dis- or relocate the soldier from the violence, thus easing some aspect of soldiering. While human soldiers certainly remain connected to the enactment of drone violence somewhere in the kill chain, their bodies can be relocated elsewhere. Symbolic then of presence and absence, visibility and invisibility, and disembodiment, drones are not technologically interesting or innovative in and of themselves but, rather, gain meaning in terms of how they supplement humans as part of a socio-technical system in an architecture of violence at a distance. As such, drones are part of a longer historical military desire to create increasing distance between the subject and object of violence in conflict. Beginning with balloons during the revolutionary wars, the initial idea was rather basic: to enact an eye in the sky and inject height and verticality into the projection of force (Adey, Whitehead, & Williams, 2013). Later, the aim was to give that same martial gaze the potential for lethality through a logistics of military perception (Bousquet, 2018; Virilio, 1989). Today, drones can be sophisticated, using artificial intelligence to identify, track, and engage targets, or they can be rather basic and cheap, designed to be disposable. They can be used for surveillance and reconnaissance, with or without precision-guided weapons attached. What matters is that, whether big or small, weaponized or not, drones work to extend the scope and scale of violence while *at the same time* reducing or eliminating the physical, cognitive, and sociopolitical burdens of contemporary war-making, in general, and soldiering, in particular. In doing so, drones are desirable because they are thought to (a) protect (some) life and (b) to enable and concentrate combat power, or *mass* in military speak. This ability to act as a force multiplier – to enhance combat potential – is especially important for casualty-averse societies (Kaempf, 2018) that deploy drones to minimize risk on their side (Renic, 2020). This risk transfer has reconfigured our understanding of the meaning of bodies and embodiment in war (McSorley, 2013; Wilcox, 2015) while raising many parallel and often intersecting critical ethical, legal, sociopolitical, and cultural questions about the place and meaning of drones for soldiering and contemporary war-making (Asaro, 2013; Chamayou, 2015; L. Clark, 2019; Enemark, 2021; Parks & Kaplan, 2017; Pugliese, 2013; Richardson, 2022 Zehfuss, 2018; Wilcox, 2017).

This book contributes to these lines of inquiry while asking the reader to consider what drones might tell us about the meaning of military *sacrifice* as it relates to the status of war. My core argument is that drones cannot be disentangled from sacrificial politics – how sacrifice is politicized, governed, displaced, and lost – namely, how it is both demanded and disavowed. However, when we inquire into militarized sacrificial politics, we bump up against a pervasive yet agitating logic of *privatization*, which I claim perverts the social and historical meaning of military sacrifice. Therefore, drones *also* cannot be disentangled from an analysis of the politics of privatization, as privatization has been key to the development and deployment of drones. Drone warfare, as evidenced in the global war on terror and its aftermath (Dauphinée & Masters, 2007; Greenwald, 2013; Pemberton & Hartung, 2008), is thus intertwined with sacrificial politics on the one hand, and a logic of privatization on the other. Together, drones, sacrifice, and privatization form an orienting triangle at the centre of this book. Each point of the triangle is related to (and contingent on) the others. Together they provide a lens through which to zoom out and view the changing meaning of bodies in war at a macro level. When zoomed in to the micro level, we find at the core of this triangle is a decaying citizen-soldier archetype. This triangle then provides an analytic framework to advance a sociological account of drone warfare. We can also better understand what is at stake given the changing quality and character of war's violence, which I will claim is increasingly *antisocial*. This antisocial quality of war has profound implications for how liberal democratic societies have traditionally understood war, and to what degree they will be able to "know," stop, prevent, and remember war in the future. Thus, drones force a reconfiguring in terms of how we can come to critique and curtail war's contemporary violence.

War's contemporary violence has an explicitly corporatized underbelly (Freeman & Minow, 2009; Wedel, 2008) as seen in the current booming business of drone-saturated violence: At least 113 countries and 65 non-state actors have weaponized drones (Patton Rogers, 2024). In Russia's current war against Ukraine,[1] Russian forces are primarily using Iranian-made Shahed-136 drones, which have a wingspan of 2.5 metres and cost just US$20,000; as well as the Orlan-10, a reconnaissance drone that can carry small bombs (BBC, 2023). Ukraine, which is reportedly losing 10,000 drones per month, mainly deploys the Turkish-made Bayraktar, which is the size of a small plane and can be armed with laser-guided bombs (Shevchenko, 2023); US-supplied Switchblade kamikaze drones; and cheap, off-the-shelf quadcopter commercial drones, like the extremely popular Mavic, manufactured by China's DJI (Franke, 2023). As anxiety continues to mount regarding the United States's future role in delivering military aid to Ukraine, on 19 February 2024, Canada announced a donation of 800 SkyRanger R70 surveillance drones (CBC, 2024), while Latvia is leading a coalition of European countries committed to producing and supplying Ukraine with "one million drones" (Breaking Defence, 10 February). Israeli Defence Forces (IDF) are using drones to infiltrate Hamas' labyrinth underground tunnels, regarded as a "soldier's nightmare," due to the physical and mental challenges of disorienting subterrain warfare. In response, the IDF is deploying Shield AI's Nova 2 drone, which uses "state of the art path planning computer vision algorithms to autonomously navigate complex subterranean and multi-story buildings without GPS, comms, or a human pilot" and thus, according to Shield AI's website, has "proven its value in combat by bringing service members home to their family." Moreover, drones feature heavily in the United States's maritime strikes against various Iranian-backed militias in the Red Sea, such as the Houthi forces in Yemen, who have successfully deployed reconnaissance and combat drones since 2015. On 18 February 2024, the United States said it struck five Houthi targets in Yemen (Barnes, 2024), including an underwater drone. Two days later, as part of a series of drone-based attacks, Houthi militias claimed responsibility for shooting down an American MQ-9 Reaper drone. Drones (both large and small) are now regarded as central to how major powers conceptualize and implement military strategy (Horowitz, 2021; Scharre, 2018). Rather than deploying troops, countries would prefer to donate weapons and drones. As a result, there is a widening industrial base to support their design, development, and deployment, which runs in parallel with a linked waning desire and ability to deploy troops.

The United States's "Replicator" program serves as an example. Announced in September 2023, it seeks to "field attritable autonomous systems at a scale of multiple thousands and in multiple domains," dramatically ramping up drone production with an emphasis on fast and efficient scaling (Garamone, 2023). The US Air Force wants "a thousand [wingman] drones to accompany their fighter pilots into combat" (Decker, 2023). Moreover, Australia announced in February 2024 plans to invest AU$400 million to develop its homegrown MQ-28A Ghost Bat autonomous armed drone program (Green, 2023). A collaboration between the Air Force and Boeing Australia, the Collaborative Combat Aircraft, a reflection of human–machine teaming, is "designed to act like a loyal wingman which will be able to protect and support our military assets and pilots and undertake a wide range of activities across large distances, including performing combat roles" (Australian Government, Defence, 2024). Likewise, in August 2023, Iranian president Raisi unveiled the Mohajer-10 drone. Like the MQ-9 Predator, manufactured in the United States, the Mohajer-10 can carry a 300-kg (about 661.39-lb) warhead and travel non-stop at an altitude of 7,000 meters (about 4.35 mi) and has the required range to hit Israel. While boasting this acquisition would alter how Iran is viewed in the region and across the world, Raisi also remarked that "yesterday they viewed us as a consumer and a country in need. Today, they see us as a producer, who can have much to say in the defence and military industries" (Motamedi, 2023). Finally, North Korea recently conducted a test of its Haeil-5–23, a nuclear-capable underwater attack drone (Shin, 2024). New "drone powers" (Patton Rogers, 2024) reveal how countries are competing to field, produce, and export emerging technologies to shape the international order.

Due to the widespread democratization of drone technology, countries that have not traditionally been seen as leaders in military affairs now have armed drones. China, for instance, is now the world leader in drone exports (Rasheed, 2023). To combat technology acquisitions in regions that are hostile to American and allied interests, particularly those who may inflict drone violence on US soil, there is now a "counter-drone" industry growing across the United States, Europe, and China (a market expected to be worth 4 billion GBP in the next decade) consisting of products to detect, track, electronically jam, destroy, or commandeer rouge drones (Holmes, 2019). Such an industry, devoted to taking down from the sky the very same drones that same industry initially helped put up, is populated with members of military intelligence services who often "transfer their knowledge to start hi-tech companies after they leave the army" (Holmes, 2019). To be sure, with every measure comes a countermeasure, and such an endless loop undoubtedly continues to spin the wheels of technology's link to militarization, benefitting industry executives, former military personnel looking for career or business opportunities, and those who otherwise stand to profit from these systems.

Rather than focusing on drones at the strategic, operational, or tactical levels, this book views the drone as a metaphor to think more broadly about war and sacrifice. In considering the social and cultural underpinning of drones, for instance, we might ask: What is it exactly that the Replicator program is attempting to replicate? One obvious and straightforward explanation might simply say it is replicating drones – as drones are lost in combat, other drones must arrive quickly to replace them. Yet something is unsatisfying about this explanation. Indeed, another view might say that it is attempting to replicate human capability absent humans. Replicator is trying to compete with China's advantages in mass, where China has more ships, more missiles, and more people with scalable, all-domain "attritable" autonomy or ADA2 (J. Clark, 2023). What does *attritable* mean? It relates to attrition, and it means systems that are disposable and can be easily replaced, systems that are cheap and inconsequential if lost, systems that do not really alter the course of battle one way or another, systems that can be regenerated and reproduced indefinitely to wear down one's enemy, and suicidal systems like "kamikaze" drones that loiter and then crash into a target and explode. As I suggested earlier, we can think about drones as extensions of human capability with the potential to overcome the politics integral to wounds, injuries, and deaths of soldiers – to override war's embodiment and "pain's inexpressibility" (Scarry, 1987). In this way, drones reconfigure the place and meaning of bodies, and thus sacrifice, in war. To that end, the Replicator program and ADA2, more specifically, view drones as sacrificial in that they are just disposable things that can be easily replaced. Yet drones can also be weaponized to execute sacrifice – to kill – from above. The power of the drone is therefore contingent. This book explores the relationship between drone warfare and ideas integral to sacrificial politics.

Drones are now a common feature of daily life for those living in "ungovernable," spaces (Akhter & Shaw, Cavallaro, Knuckey, & Sonnenberg, 2012; Gregory, 2020; Sluka, 2013) where, according to Derek Gregory (2017, p. 29) "a particular group of people is knowingly and deliberately exposed to death through the political-juridical removal of legal protections and affordances that would otherwise be available to them." At the site of the botched raid in Yemen, and indeed across multiple target zones, the sound of drones buzzing had become a recognizable one. According to Michael Richardson (2022) "the insistent, unceasing noise eats into the body, working at the affective level of embodiment 'to the extent that the sound of loitering drones triggers mental and bodily responses – indicative of post-traumatic stress disorder – in advance of a drone strike having actually taken place" (Schuppli, quoted in Richardson, 2022, p. 43). On that evening, however, the drones were flying lower than usual. "Though U.S. boots have been on the ground in Yemen on and off since 2002, drones and manned jets lead the hunt for AQAP [Al Qaeda in the Arab Peninsula]. More than 182 strikes have left 815 people dead, including 134 civilians" (Al Sane & Shabibi, 2017). Yemen is merely one example of a nation that has become the target of the US-led drone program – a space of exception. The first strike there was devastating: "Commanders thought they were targeting al Qaeda but instead hit a tribe with cluster munitions, killing fifty-five people. Twenty-one were children – ten of them under five. Twelve were women, five of them pregnant" (Purkiss & Serle, 2017). According to the human rights organization Reprieve, in Pakistan, Somalia, Libya, Iraq, and Afghanistan, drones have executed, without trial, some 4,700 people (Reprieve, 2017). Airwars, a not-for-profit transparency watchdog that tracks, assesses, archives, and investigates civilian casualties across eight air-based conflicts, has assessed 62,509 alleged civilian deaths since it began tracking in 2014. In Yemen, it estimates between 78 and 159 civilian deaths across the thirteen-year campaign, while the United States has confirmed just 13 civilian deaths. There were more strikes in President Obama's first year as president than there were in the entirety of President Bush's time in office – a total of 563 strikes, largely by drones, were carried out during Obama's two terms compared to a total of 57 strikes under Bush (Purkiss & Serle, 2017). Another estimate by Micah Zenko (2017) puts it at 542 drone strikes, which killed roughly 3,797 people, including 324 civilians. As former US president Barak Obama, the Nobel Peace Prize winner in 2009, reportedly told senior aids in 2011: "Turns out I'm really good at killing people. Didn't know that was going to be a strong suit of mine" (January 20, 2017). According to Ryan Cooper (December 1, 2021), the Trump administration decentralized control over targetting, rolled back the already modest protections for civilians, and dramatically increased air strikes in 2017. That year also saw Trump stop disclosing significant information about US troop deployments in Iraq and Syria, even while the military was taking on a greater role there (*Los Angeles Times*, March 30). During Trump's first term, "air and artillery strikes in Iraq and Syria created more than twice the number of casualties compared to Obama's second term. … In Somalia, casualties increased roughly eightfold" (Cooper, 2021). The Biden administration, according to a classified new policy, departs from the Trump administration in that it required President Biden's approval to add suspected terrorists to the "kill list" (Savage, 2022). The war on terror spanned multiple administrations and witnessed the unacknowledged loss of many civilian lives. Along the way, the United States became exceptionally good at killing people in a way that signalled to the world that not all deaths in the global war on terror would count (Cloud, 2011), that not all deaths would be recognized or considered equally meaningful.

Most importantly, for our purposes, this inequality would be made bare in that not all deaths would be considered sacrifices.

To this end, another part of the story of the botched raid is yet to be told, which is equally as striking in relation to the centrality of drones and death wrought by disembodied air power from afar. It concerns the sole American casualty – Navy SEAL William "Ryan" Owens. Consider some of the statements made by US officials about his death following the raid. Former United States Press Secretary Sean Spicer said the raid was "absolutely a success, and I think anyone who would suggest it's not a success does [a] disservice to the life of Chief [Petty Officer] Ryan Owens" (Pengelly, 2017). Another White House spokesperson, Michael Short, stated that Ryan Owens was "an American hero who made the ultimate sacrifice in the service of his country" (Pengelly, 2017). Deputy press secretary Sarah Huckabee Sanders said, "As a parent, I can't imagine the loss that [Ryan's father – Bill Owens] has suffered. I think every American owes his son a great deal of gratitude. We are forever in his son's debt. I know that he paid the ultimate sacrifice when he went on that mission" (Pengelly, 2017).

Not everyone shared in this positive construction of events. Bill Owens refused to meet with the president upon the return of Ryan's body to US soil, claiming, "Trump should not hide behind my son's death to prevent an investigation" (J. Brown, 2017). In a similarly heartbreaking moment ten years prior, Cindy Sheehan, a prominent antiwar activist and mother of fallen soldier Casey Sheehan,[2] would also point to the politicized use of sacrificial language. In resistance to the state's attempt to incorporate Casey's death into a narrative of sacrifice, she frequently talked about how her son was used as sacrificial cannon fodder. A decade later, President Trump, in his first address to Congress on 28 February 2017, said of Ryan's legacy that it would be "etched into eternity" (Baker, Hirschfield Davis, & Sheer, 2017). Still, Bill Owens refused to meet with the president. He called it a "stupid mission," stating that "for two years prior, there were no boots on the ground in Yemen – everything was missiles and drones – because there was not a target worth one American life. Now, all of a sudden we had to make this grand display" (Pengelly, 2017).

"A grand display" – the presence of US troops as fleshy and vulnerable beings on the ground. Where combat once involved conventional troop deployments and thus an element of risk, seeing soldiers' bodies be made invisible, protected, and remote is now normal to preserve American life at all costs. Although every war requires the making of human killing machines, today it seems as though soldiers need not go to war expecting to die but only to kill (Asad, 2007). According to Bill

Owens, not one American life was worth sacrificing. A grand display amounts to what would have historically been considered a familiar outcome of war: embodied and violent, albeit potentially unplanned and/or undesired, encounters with an adversary. For Bill Owens, this "display" of bodies, their mere physical presence in a war theatre, amounted to an unfortunate anachronism. *What was the point of it all,* we might infer. Why bother with troops in war when advanced remote military technology can make conventional deployments unnecessary? Yet, for Sean Spicer and the others, this loss *had a purpose.* We might suggest that it, among other things, justified the existence of, or helped to reconstruct, the nation (Anderson, 1991). The result of the administration's position, although outwardly cloaked in love and gratitude, amounts to the normalization of martial violence in defence of the nation and, as a result, the silencing of criticism or alternative explanations for soldiers' deaths. How did Sean Spicer and others achieve this double move, transforming death into a noble cause, while normalizing martial violence? They did so, I suggest, through the alluring, transcendental theme of sacrifice (Nancy, 1991).

Military sacrifice is frequently an assumed given in war. It is a seemingly natural and glorified act that binds citizens to nation-states in a traditionally reciprocal relationship, sometimes described in reference to a social contract, where citizen soldiers give to the state in return for certain rights and privileges. In the case cited earlier, US officials managed to depoliticize the violent event of the raid, rendering the death of Ryan – that which "we are all forever indebted to" – *outside of politics.* In labelling the death a sacrifice, it becomes transcendental, and the political dimensions are foreclosed. In giving it the gloss of sacrifice, the death was translated into an untouchable otherworldly event, one simultaneously comprehensible and incomprehensible. In deploying sacrificial language, they render the raid, and the deaths of Yemeni civilians, including children, morally and politically justified by extension. Sacrifice is an assumed given in war, and its character is both political and non-political. In this sense, sacrifice normalizes nation-states (Appadurai, 1996), giving us both a structure and a language to enable a myth made up of constituent units (Lévi-Strauss, 1955).

The myth of sacrifice was utilized to, as Bill Owens so acutely observed, "hide" the material, antagonistic politics of war's senseless, brutal, and unjustified violence. Sacrifice, then, offers a linguistic and narrative glue by which order can stick in moments of violent disorder. In these contested moments of violent aftermaths, fallen or wounded soldiers emerge as symbols for how nation-states manage the meaning of life and death in wartime:

> [T]he indeterminate status of the recently dead is a source of great anxiety because death threatens the identity of the living to whom the deceased was bound. Proper words and gestures – even angry ones – are a means of responding appropriately to this threat. They serve – in the funerary rites and later – to incorporate death into the predictable continuity of a form of life and thereby to suppress the thought that it is life that is contingent. Thus, it is not the occurrence of death as such in which horror resides but the manner in which it occurs and how the dead body is dealt with by the living. (Asad, 2007, p. 78)

Military commanders are bound to their soldiers. Soldiers' deaths, to mitigate great anxiety, must *have a purpose*. To interrogate Ryan's death, for example, by refusing the dominant political narrative as Bill Owens did, would be to, according to the statements made by the White House representatives, belittle the life lost. Despite their best efforts, Bill Owens's words and gestures, and Cindy Sheehan's words and gestures before him, respond to the failed attempt of these officials to translate death into a sacrifice. Here, we see the failure of sacrifice to take hold, to be accepted. The idea that military sacrifice is a universal feature of war is questioned in these moments of failure. Sometimes, military deaths are simply unjust and have no agreed wider meaning. Sometimes, sacrifice fails. In moments of failure, we can more clearly see the politics of sacrifice and the need therein to quell the anxiety of the dead's indeterminate status.

This book is about how, on the part of citizen-soldiers, military sacrifice, in its material, symbolic, and discursive forms, both fails and survives in contemporary wars. We often assume that war *needs* sacrifice; at a very basic level, this book asks whether this is true and what the implications might be either way. This assumption about war's need for military sacrifice is not unfounded but comes from an inherited tradition of French revolutionary wars, and a subsequent Clausewitzian military doctrine, whereby combat is necessarily a physical encounter, which has been well socialized across time and place over a period of more than 200 years. Therein, war became tied to the archetype of the citizen-soldier, whose sacrifice for the nation and in defence of the sovereign (represented now by the people) was normalized, glorified, and universally honoured, culminating in the Great and Total Wars of the twentieth century and the subsequent social contract based on rights and responsibilities of states and citizens. Without much debate or curiosity, we all universally pay our respects to soldiers who fought and died in these just wars. But what if future wars did not need soldiers to sacrifice? What if war carried on in the absence of embodied encounters with adversaries, eclipsing the injuries, wounds, loss, and politics associated with citizen soldier sacrifice? Would it matter, and if so, in what ways?

Here you may be thinking: Is sacrifice not something that we want to avoid anyway; *is it not bad?* Are we not all just trying to make war better? Is the preservation of life at all costs – indeed, the avoidance of sacrifice – *not a good thing?* Is making war a more humane practice, with fewer casualties, *also not a good thing?* Putting these questions aside for now, consider a scenario in which war overcame injuries, wounds, loss, and the associated politics and was instead totally bloodless on "our side," driven by the need to avoid, deny, or conceal sacrifice. Let me first swiftly counter one possible thought: This book does not argue that we have entered a technocentric post-sacrificial period of war. One need only look to Ukraine, Palestine, Lebanon, Sudan, and elsewhere to see that war's materiality is real and deadly. Rather, it argues instead that liberal states, particularly the United States, the focus of this book, *face a problem* in waging and sustaining "wars of choice," namely, how to manage the tension between life, something that we ostensibly universally value, and death, something that we wish to avoid but which nevertheless happens in war. To manage this tension, liberal war-making paradoxically proclaims a desire for peace while simultaneously increasing military capacity for war (Reid, 2006). In this quest for peace under militarization (Stavrianakis, 2016), military sacrifice, I argue, comes to be explained and rationalized through governing logic. This governing logic means that states must be positioned to deny sacrifice while still maintaining the grounds to demand sacrifice. Or, to use the language of Michel Foucault: to balance the contradictions between sovereign and biopolitical power (Edkins, Pin-Fat, & Shapiro, 2004; Mutimer, 2005; Rabinow, 2006). In North America and elsewhere, military recruitment is waning (Della Volpe, 2015; Risen, 2025; Turnbull, 2024), and deploying conventional forces is politically unpopular. Yet war's violence, to be legitimate, must be able to harness the myth of sacrifice in relation to soldiers and, on occasion and where necessary, summon a romanticized past in which sacrifice is an ordinary feature of war, albeit contested and uneven. How do states manage this balancing act? How can states both demand and deny sacrifice?

I argue that this balancing act is managed through the governing of military sacrifice, drawing on military privatization and drones as two interrelated examples: privatization and drones are two sides of the same coin. Both signal some form of departure of the citizen-soldier from the battlefield and thus culminate in an absence even as the presence and lethality of combat power intensifies and concentrates. The initial step in the governing of military sacrifice was to outsource and privatize war. I claim that drones are the technological expression of the political philosophy that underwrites military privatization. Both work *together* to destabilize the citizen-soldier's relation to sacrificial cults and idioms. The events of the French Revolution, two World Wars, and the ideals of liberal democracy as they confront proxy, limited, and/or counterinsurgency wars, together signal a legacy in our thinking about *what war is.* But these are historically specific events and ideals. There are no guarantees that war remains the domain of citizens and citizen-soldiers. Clausewitz's view of combat is no longer widely applicable.

Asking that war remain tied to a sense of embodiment is a way to ensure that it be both rare *and* political, because wounds are political (Enloe, 2019). Without concern for or knowledge of bodies in war, where can we locate the politics or ethics of war? This is why the rhetorical questions stated earlier, about war becoming more technologically precise and thus ostensibly more humane (Moyn, 2021) and "better," (Arkin, 2010) while perhaps well intentioned, miss the point: The point is to ask what is lost when (some) humans are deliberately made absent in war. Sacrifice allows us to drill down into this question. The goal here is to interrogate what sacrifice, whether concealed or amplified, may tell us about war's nature and character and the future of war more broadly. Governing military sacrifice, I argue, is about managing the politics and contradictions related to how soldiers can go to war expecting to kill, but not die (Asad, 2007).

Why does this matter? It matters because, should war's violence be stripped from the body of the citizen-soldier, should war lose the need for wounds, injury, and loss, we may simultaneously lose our social, legal, and political mechanisms to enable us to know, stop, limit, critique, or prevent war, because doing these important things has historically required that bodies be central to the narratives of war memory and transgenerational knowledge about war's harms, as well as demilitarization, and or anti-war movements. Both privatization and drones, when taken together as explanations for the decline of citizen-soldiering's link to sacrifice, suggest trends not only towards disembodied war but *depoliticized* war. Exploring military sacrifice as something that is *not given* in war but, rather, is a political outcome, as this book does, allows us to think about the future of war where sacrifice might not be relevant. Using sacrificial language today often reveals wartime anachronisms that contradict how war-making is actually practised in liberal societies. This book attempts to draw attention to these anachronisms, asking what productive function they may serve, and what is at stake in a future in which war does not require sacrifice – when war risks becoming antisocial.

Methodological Approach

This book asserts a paradox integral to military sacrifice today: How do liberal governments both demand and deny sacrifice? My aim is to contribute to postmodern approaches in critical security studies that build on the "discursive turn" in international relations. As a sociologist, I am naturally inspired by C. W. Mill's (1959) "sociological imagination." Mills claims that "neither the life of an individual nor the history of a society can be understood without understanding both" (p. 3). In this way, the figure of the "citizen-soldier" is a centre piece of this book. To better understand this figure and how privatization and drones have impacted it in theory and practice, I conducted twenty interviews with security and defence professionals, military veterans, and anti-war activists. Interviews thus fell broadly into two camps of people: practitioners and critics. Practitioners consist of security and defence experts, drone lobbyists, and United States military veterans who transitioned to the private sector in drone consulting, research and development fields. Critics include US military veterans, anti-war activists, and two military veteran whistleblowers who were deployed to Iraq and Afghanistan. All interviewees had a direct role in the application, conceptualization, and/or practices associated with military drones. As I show, practitioners drew on depoliticized, distanced, and sanitized abstraction and euphemism, which I suggest serves in the stabilization of their subject position as agents, experts, and actors, rather than as passive victims of drones. Consider the phrases "clean bombs" and "surgically clean strikes," which Carol Cohn found circulated in nuclear defence discourses in the late 1980s. Such phrasing demonstrates "an astounding chasm between image and reality that characterizes technostrategic language" (Cohn, 1987, p. 692), reminding us that the discursive deployment of "clean bombs" and "surgically clean strikes" has a historical lineage that predates our contemporary technological moment of drone warfare. Not much has changed. In contrast to the euphemistic and sanitized abstraction characteristic of the security and defence security experts, critics portrayed drone warfare as fraught with intimacy rather than distance. Many interviews encountered poverty or poor job prospects prior to their enrolment in the military; many have been diagnosed with post-traumatic stress disorder, traumatic brain injuries, and other mental health disorders following their military service. Whereas practitioners discussed drones with the language of economic efficiency and clean, humanitarian types of warfare – language choices often deeply intertwined with euphemisms of neoliberal free-market capitalism (Harvey, 2005) – critics drew from their experiences of trauma and suffering in a way that linked the personal with the political and thus implicitly invoked Mills's sociological imagination and ideas pertinent to the citizen-soldier archetype therein.

Understood here as a significant archetype which binds war to citizens in the public sphere within democratic states, the citizen-soldier archetype cannot be separated from its entanglements with shifting citizenship regimes, globalization, technology, and economics related to the production of military violence writ large (Arrighi, 2007). This book analyses the citizen-soldier in relation to the dialectics of demanding and disavowing sacrifice. In this way, I advance Taussig-Rubbo's (2009) claim that sacrifice orders violence within a schematic of significance and insignificance. A key argument of this book is that the solidification of the relation between citizenship, soldiering, and sacrifice has historically been marked through a schematic of significance and insignificance: Whose deaths are noble? As an archetype, the citizen-soldier reflected the highest echelon of sacrifice and a turning point in the legitimacy of nation-state violence. Yet, that archetype is increasingly troubled. Of importance here for historicizing the citizen-soldier's link to sacrificial cults and putting into context our contemporary moment are the critical turning points of the First and Second World Wars (Total wars) and, likewise, the political and ideological intersections between technology and militarization after the Vietnam War and the first Gulf war in 1991. Let us briefly trace the changing status of the citizen-soldier across the twentieth century.

The First World War was a gruesome outcome of combining new weapons technology with inexperienced troops and an inadequate command and control culture. It reflected "the combination of modern nationalism, industrialism and technological limitations … whose defining characteristic was mass" (Shimko, 2010, p. 10). The inaccuracy of the newly acquired firepower used in First World War was eventually resolved during the nuclear revolution of Second World War, which revealed the "speed with which total annihilation [could be] carried out, and the ability to do so without first achieving success on the battlefield" (Lieber, 2005, p. 126). From machine guns to air assaults and bombing, the Second World War altered how technology would be conceptualized within the spatial logic of the battlefield. Accordingly, American war-making in the Second World War and into the Cold War period was defined by a military doctrine of annihilation along with a resource-based approach to warfare, as evidenced by the conflicts in Korea and Vietnam (Adamsky, 2010, p. 79). While more than 19 million military personnel died in combat during the Second World War, the deaths of allied soldiers were regarded as just and honourable and therefore are remembered as noble sacrifices. In "wars of necessity," citizen-soldier deaths were considered legitimate. Sacrificial politics were consolidated into the nation's imagination of itself during this time. In contrast, limited wars, such as humanitarian wars or "wars of choice," turn casualties, and therefore the problem of how to incorporate sacrifice in a schematic of significance and insignificance, into a political problem to be managed (Mandel, 2004).

In interrogating this problem, I join others who seek to productively challenge the field of international relations and its tendency towards realism and associated positivist methodologies, as well as its preoccupation with viewing sovereignty, the state, violence, and war as inherently bounded, knowable, progressive, self-evident, and unit-based occurrences. These critical interventions are characterized by Foucauldian impetus towards what he termed "archaeological" and "genealogical" methods as well as deconstructive methods (Mutlu & Salter, 2013). Together, these methods constitute a general suspicion of modern metanarratives and a congruent trend towards a historicity, closure, and essentialism, as well as the belief that transcendental constructs, such as reason and consciousness, are historically and culturally contingent (Fournier, 2014). Briefly, the discursive approach in critical security studies theorizes that "language is political, social, and cultural: Discourse analysis is the rigorous study of writing, speech, and other communicative events to understand these political, social, and cultural dynamics" (Mutlu & Salter, 2013, p. 113). The discursive turn refers to the methodologically broad ways in which social meaning is derived from the interplay between various textual artefacts across time and space. For critical security studies' topics and approaches, where genealogy, intertextuality, and speech acts make up the primary concepts, discourse analysis has as its object of study the sociopolitical world: continuity, change, and rupture are its strategies (Mutlu & Salter, 2013).

For example, official discourses of the "exceptional" war on terror are regarded here as deeply connected to other more "routine" politicized speech acts that have war, sacrificial violence, memorialization, and militarization as their core elements. In this way, discourses are not discretely bounded entities but must be contextualized in relationship to discourses past and present. "Discourse is not produced without context and cannot be understood without taking context into consideration ... discourses are always connected to other discourses which are produced earlier, as well as those which are produced synchronically and subsequently" (Fairclough & Wodak, quoted in Phillips & Hardy, 2002, p. 4). Discourses reveal a historical and relational quality that permits the ascendance of some ideas over others according to a relation of knowledge and power; a relation that the social construction of ideas cannot escape. Social structures not only determine discourse, but they are also a product of discourse (Fairclough, 1989). In this way, the theme of exceptionalism contained within the discourse of the war on terror, and the routine speech acts that appear more banal in the everyday, are co-constituting.

The global war on terror saw the geopolitical expansion of war powers but its architects were famously casualty-averse. Over time, we have seen the emergence of war strategies concerned with the importance of force multiplication and efficiency linked with the concern to reduce casualties and remove troops from battle such that casualty minimization is now the yardstick for success (Mandel, 2004, p. 1). Following the Second World War, and throughout the Cold War, US casualty aversion escalated, culminating in the post-Vietnam era. Casualty aversion escalates in moments when the public questions the legitimacy of wars (Shimko, 2010) and seems to reflect historical qualities of American political culture, for example, the value of the individual over the state and a historically deep-seated suspicion of government (Mandel, 2004). Casualty aversion is one key aspect of the war on terror. Another important legal development is the Authorization for the Use of Military Force (AUMF).

The AUMF has been re-authorized with every US administration into the present. Originally signed into law by President George W. Bush on 18 September 2001, it means that any entity thought to be associated with al Qaeda or associated forces and/or perceived past or future violence towards the United States may be subjected to American military power without the president having to consult Congress. With unprecedented scale and breadth, this law signalled the potential for forms of state violence – so-called force short of war – without end, temporally, spatially, and politically. But this widening of the scope of violence requires the ability to call for sacrifice. At any point, soldiers may be called on to satisfy the multiple and ongoing aims of the war on terror (even as it appears to have ended) across the multiple geopolitical locations in which it unfolds. The war on terror, therefore, includes moments where sacrifice is considered undesirable and as something to be avoided, as is the case with the ideology of casualty aversion. It also includes moments where sacrifice is openly needed, acknowledged, and celebrated, as the AUMF would support. This book suggests that we ought to pay attention to both the demand and denial of sacrifice to better comprehend the current and future status of war.

As America's longest war, including both declared and undeclared conflicts, the war on terror marked a shift from costly ground wars to a focus on generating greater physical distance between the subjects and objects of violence primarily through forms of unmanned air power (Coeckelbergh, 2013). Consider Afghanistan, where air attacks dramatically increased while President Obama withdrew most troops at the end of 2014. The key point here is that the removal of *some* bodies from a conflict zone does not correlate with a decline in militarized violence but instead with its restructuring. As citizen-soldiers are pulled from conflict zones, often they are replaced with private military personnel. In the United States, these deployments are not overseen by Congress, and their deaths are not publicized (Avant & Sigelman, 2010; Kinsey, 2008; Maogoto, Newell, & Sheehy, 2009). They also happen to be key partners in the CIA's counterterrorism and drone programs (Coll, 2014). At the height of the war on terror, the company formerly known as Blackwater was active in the assembly and loading of Hellfire missiles onto Predator aircraft, work formerly done by CIA personnel. Benjamin (2013) quotes war correspondent Jeremy Scahill, claiming, "it is Blackwater that runs the program for both the CIA and JSOC [Joint Special Operations Command] because contractors are not [overseen by Congress] so they just don't care. If there's one person they're going after and there's thirty-four people in the building, thirty-five people are going to die. That's the mentality" (pp. 63–64). Private military contractors offer militaries and states forms of physical labour, on one hand, and technical or logistical expertise, on the other. In wars on terror, which find their origins in casualty-averse societies, private military corporations are but one symptom of the governing of military sacrifice.

We may look to the revolution in military affairs (RMA; Beier, 2012) as spurring the inclusion of private military contractors into war-making practices. Central to the RMA is net-centric warfare, the aim of which is to link up a smaller number of highly trained humans with small, fast, agile weapons systems and mechanized support linked via GPS and satellite communications into an intricate, interconnected system in which the behaviour of components would be mutually enhanced by constant exchange of real-time battlefield information (Eco, quoted in Lucas, 2010, p. 290). Net-centric warfare produces and sustains the technological conditions necessary to achieve bodyless warfare through drones and roboticized models of war fighting, as machines can operate in hazardous environments, do not need extensive education and training, require no minimum hygienic standards, do not tire or let emotions cloud judgment, and are perfect in suppressing the enemy since soldiers do not even have to be exposed to an enemy in the first place (Eco, quoted in Lucas, 2010, pp. 363–364). The RMA thus facilitates the desire for full-spectrum dominance and surveillance of land, air, and sea, the militarization of space, information warfare, and control over communication networks. This also entails the expansion of military bases globally. Yet, instead of boots on the ground, operations are now conducted by fewer people, such as technicians in air-conditioned tents (Parenti, 2007) and/or private military contractors who fill gaps in information collection, management, and analysis required by the objectives of netcentric war.

Significant force reduction is required to finance the technology necessary for the RMA (Moskos, Williams, & Segal, 2000, p. 5). The military was not exempt from broader institutional transformations in industrial societies: it too shifted from a labour- to a capital-intensive organization (Manigart, 2006). The decline of the mass army model went together with restructuring towards professionalization. Managers and technicians increasingly conducted war, as opposed to combat leaders (Moskos, 2000, p. 15). The professionalization of the military was essential for the realization of the goals of the RMA: "a policy agenda emphasizing the exploitation of technological advances to preserve and even improve the United States' long-term strategic position" (Shimko, 2010, p. 2). Without professionalization, the ideals of the RMA, particularly incorporating precision-guided weapons to give a qualitative advantage to the United States over the quantitative advantage of the Soviet forces, would have remained abstract (Adamsky, 2010, pp. 59–61). The professionalization of the military for purposes of net-centric warfare included a parallel turn towards the expert knowledge offered by private military corporations and remote military technologies.

A cursory glance at international politics shows a growing trend towards technological means to enable a vision of risk-averse, disembodied combat. Combat "unmanning" includes various ways to talk about the same thing: unmanned aerial systems, unmanned aerial vehicles, remotely piloted aircrafts, or the preferred shorthand "drones." As stated earlier, these systems are all about combining efficiency with the need to protect troops, and they are now prolific. Surveillance and weaponized drones are now a mainstay of any modern or modernizing military and are featured heavily in the current Russian–Ukraine conflict and Israel's war against Hamas; they are no longer new or controversial. Drones are a pillar of the United States's counterterrorism and counterinsurgency operations and their centrality in how states imagine current and future security environments is only increasing. In the United States, the synchronistic efforts of the US government, the Pentagon, arms manufacturers, the CIA, the Federal Bureau of Investigation, the JSOC, and private military/security corporations to develop and integrate drones and other emerging technologies underscore a quest for distanced, riskless, and bloodless war. Drones and other related means of remote war extend this ongoing quest, delivering enhanced battlefield capabilities, greater situational awareness, and perhaps most importantly, a more certain means of casualty avoidance for citizen-soldiers who, ideally, would be spared the scene of violence altogether, thus avoiding its attendant risks.

As I argue in this book, the archetype of the citizen-soldier, which emerged within the tradition of democratic republicanism, is integral to the politics and narrative of sacrifice. For sacrifice to be cognitively and emotionally legible today in the context of US-driven war-making, it must not only convey "proper words and gestures," such that the living might make peace with the death in question, but it must also hinge on the deployment of a robust conception, normalization, and acceptance of the archetype of the citizen-soldier. Yet this archetype is a product of historical circumstances, which renders its status fraught. In the following, I outline how the citizen-soldier emerged alongside ideas of the nation in both liberal and republican traditions. I argue that both "flexible citizenship" and the image and actions of the martyr together destabilize both republican and liberal theorizing on the meaning of the citizen-soldier. I then use flexible citizenship and the notion of the martyr to illustrate the decay of the citizen-soldier's link to sacrificial cults. But first, we must briefly sketch some important concepts that will be repeated throughout this book.

Key Concepts

How is military sacrifice made possible? The assumed presence of sacrifice in a nationalized imagining of the state's monopoly of military violence, its *normalization*, is what makes it especially ripe for critical social inquiry. Given that state power is evidenced by its ability to monopolize the communicative potentiality of sacrifice (Taussig-Rubbo, 2009) to take it as a site of political struggle is to think critically about how liberal states rationalize wartime killing. This book extends Foucault's conceptual work on governmentality, sovereignty or sovereign power, and biopolitics or bio-power. In the following is a simple definition of each. However, it is important to consider their interrelation across a genealogical, non-linear view of history, and their co-evolution considering Foucault's approach to power. For Foucault, power is not one thing. It cannot be possessed. It is not top-down. Power is relational, heterogeneous, multi-directional, and omnipresent. Following a genealogical approach, we can trace the "conditions of possibility" that give rise to military sacrifice. While military sacrifice is associated with sovereign power, it nonetheless survives within biopolitics.

According to Deborah Brock (2019), it is through sovereign power that Foucault comes closest to a traditional understanding of power. Sovereign power is a repressive and prohibitive form of power, including "spectacles of control such as public executions to command obedience as well as the strict imposition of limits on daily life" (pp. 15–16). Its preeminent form of expression is therefore the negative destruction of bodies and things and the execution of wrongdoers. As Brock summarizes, sovereign power was first exercised by "monarchical authorities to prohibit descent, punish the non-compliant, and repress their subjects. Power, therefore, was enacted primarily through the threat of punishment" (p. 15) and the right to "*take* life or *let* live" (Foucault, 1978, p. 136, emphasis original). In contrast to sovereign power's emphasis on erasing life, biopolitical power has the health of populations as objects of regulation as its target. It is about supporting and administering life through "disease control; detailed record-keeping of births, illnesses, deaths, and so on; and new statistical methods that allowed for patterns within populations to be identified, understood, and addressed" (Brock, 2019, p. 17). Biopolitical power is therefore productive rather than negative, as it is concerned "to *foster* life, or *disallow* it to the point of death" (Foucault, 1978, p. 138, emphasis original). Biopower has neither displaced nor surpassed sovereign power but rather has been redistributed within a field of governmentality, shaping how we are governed, how we self-govern, and how we govern others (Dean, 1999). Governmentality includes non-state and supra-state entities in that it operates through "policies and departments, through managerial and bureaucratic institutions ... and through forms of state power, although not exclusively" (Butler, 2004 p. 52). It thus operates diffusely through state and non-state institutions and discourses that are legitimated neither by direct elections nor through established authority" (Butler, 2004, p. 52). Governmentality has distance built into it. The "art of governing" requires not that the governor "have a sting – that is to say, a weapon of killing, a sword – in order to exercise his power; he must have patience rather than wrath" (Foucault, cited in Burchell et al., 1991, p. 96). Governmental power is not a top-down force, nor is it "exercised through coercion but through the production of *knowledge and its claims* [emphasis added], the assertion of 'truths,' the making of meaning, and the shaping of conduct" (Brock, 2019, p. 21). Knowledge and control of life enable liberal peace at the same time as they afford liberal war (Reid, 2006). Sovereign and biopower in the context of governmentality together illustrate how liberal modernity regards the problem of life as something to be managed. Liberal government is concerned with satisfying the conditions for these two competing types of power. But how can a liberal society value life, yet have robust industries, institutions, and policies devoted to killing? Part of the answer to this can be found in the ideology and practice of military sacrifice. When we analyse sacrifice *in relation* to the archetype of the citizen-soldier, we can see elements of both sovereign and biopolitical power; their successful management can help ease the contradiction cited earlier. Of importance is how citizenship can suggest truths about whose bodies are worthy of sacrifice and what actions constitute sacrifice. This becomes especially pertinent in contemporary times of war, mourning, and loss, when the biopolitical ordering of life and death is reconfigured.

Sacrifice is critical in our understanding of martial violence, because this allows the nation to breathe life back into those who it claims perished on its behalf. This is precisely what Sean Spicer and others attempted to do with Ryan Owens's death; and what Cindy Sheehan rejected in relation to her son Casey's death. Sacrifice is an important feature of politics; thus, liberal government is deeply concerned with establishing some sense of monopoly over its discursive trajectories. This monopoly is challenged by contemporary war technologies and practices that results in a contestation over the categorization and classification of sacrifice as it pertains to the lives and deaths of citizen-soldiers and military contractors alike. This affects the biopolitical ordering of the boundaries between citizenship (the right to claim a certain kind of life) and sacrifice (the right to claim a certain kind of death). Since the domain of sacrifice is centred on killing and dying, it ultimately resides with sovereignty. However, sacrifice survives within biopolitics insofar as it needs some concept of life to be able to legitimately call for death. Moreover, as I show in the next section, historically, citizen-soldiers, in their potential sacrifice for the good of the nation, embody a biopolitical emphasis on life. Their deaths give life to other citizens and therefore the nation. Citizen-soldiers' sacrifice therefore embodies the synthesis of both sovereign and biopolitical power. In the next section, I illustrate the emergence of the citizen-soldier, followed by two challenges to this foundational archetype.

The Citizen-Soldier: A Lasting Archetype?

The schematic of militarized suffering and loss upheld by sacrificial cults is tied to the nationalized disciplining of citizen soldiering. As Foucault writes, war is no longer strictly aimed at securing territory but, instead, is enacted with populations in mind. Prior to the French Revolution, war was a private and dynastic affair that rarely intentionally involved ordinary people. Now, war is no longer for the benefit of a few powerful monarchs but for *everyone, in the name of life necessity.* The citizen-soldier is not simply the embodiment of a nation-state's capacity for violence, which demands the ultimate sacrifice of its citizens during war but of *everyone's* capacity for violence in the service of the nation. Modern iterations of citizenship, which hinge on the concept of everyone, and the importance of sacrifice to this end, is informed by the continuing legacy of the French Revolution and its radical displacement of sovereign power from the body of the monarch to the People.

The French Revolution, through this gradual and uneven transfer of power, introduced democracy, reconfiguring the relationship between the sovereign and citizens and giving birth to the concept of the People. Since then, soldier-citizenship has been marked by a willingness to potentially die for the nation and its causes. The Napoleonic Wars (1803–15) saw unprecedented bloodshed and up to seven million deaths. In the First World War, the Battle of the Somme alone (1 July–18 November 1916) saw one million people killed or injured in what would be the deadliest battle in human history. In the Second World War, there were between 35 and 60 million deaths, a debated yet extraordinary figure that is difficult to comprehend by today's standards. To recognize notable war deaths and casualties, war medals honour soldiers according to the nature of their death or injury, the duration of their service, and/or what kinds of risks they took on the battlefield. These sacrifices were considered just and were made for the People to flourish. Dying for one's patria is a notion that only emerges with the civic republican tradition and its democratic culmination in the mass draft of the French Revolution. We have inherited a tradition of war-making that has death and injury as integral to the nation's survivability and where noble death and injury give war its sacrificial character.

Yet scripting a nation into being requires more than just standing armies and the deployment of symbolic capital around the figure of the soldier: It requires deliberate effort and in turn (at times) unconscious consent to violence. As Antonio Gramsci (1950/1971) argues, the state has a profound role in the regulation of society, the production of culture and ideas, and the construction of citizenship. For example, Gramsci's concept of the ethical state reflects how the state "educates" consent on behalf of the masses through coercive and disciplinary forces in addition to hegemonic processes. Gramsci states that "every state is ethical in as much as one of its most important functions is to raise the great mass of the population to a particular cultural and moral level, a level (or type) which corresponds to the needs of the productive forces of development, and hence to the interests of the ruling classes" (p. 258). Moreover, "there is no polity without a homeland and no homeland without a founding narrative. This space is sanctified by the appearance of the sacred, which is preserved in memory by the national narrative" (Kahn, 2011, p. 154). National and foundational narratives join the People and the state in instituting a regime of rights based on inclusion and exclusion, or citizenship.

Citizenship rights have a territorial quality. According to Ulrich Beck (2000), "the national state is a territorial state: that is, its power is grounded upon attachment to a particular place upon control over membership, current legislation, border defense, and so on" (p. 4). State policies like citizenship and border security not only secure the nation-state from outside interference, but they also solidify loyalty to the state as seen in acts such as dying or killing for one's homeland. In other words, states' expressions of sovereign power have historically entailed some application of territorialization. Yet the ascendance of the liberal state has witnessed trends towards denationalization. As the war on terror has shown, the enemy is no longer a single country and/or discrete populations within. In this respect, the "national" is thought to be intimately bound up in both the "local" and the "global" (Sassen, 2006, 2009).

The territorial quality of nation-states and populations therein, which fixes identity to territory, means that the concept of the nation runs the risk of idealizing itself as an ethnically pure entity. But this falsehood of, first, ethnic purity, and second, spatial constraint, is revealed by the highly mobile and diverse categories of the "stranger," "outsider," the "immigrant" or "refugee," which together reveal the politicization (and violence) involved in symbolically or materially turning, through inclusive and exclusive forms of racism, a racially heterogeneous nation into a homogenous one in the eyes of its beholders. Nationalist imagery reinforces a false nexus between territory and identity, thus people marked as "populations" are in turn marked with ethnicity as a brute social fact, reflecting a crude essentialism and primitive rigidity which views ethnicity as an antagonizing trait to be controlled (Du Gay & Hall, 2011). Nationalism, and nationalist imagery, entail a violent and contradictory relationship with "others."

For the nation to acquire a monopoly on the legitimate use of violence, it depended in part on the application of an emerging administrative and technocratic rationality to populations. This rationality was applied to questions of populations (an emerging site of social and political inquiry) and their security, but it was also applied to state narratives of violence and sacrifice, which created the citizen-soldier and codified the archetype's significance. Nationalization, territorialization, and colonization, which developed notions of healthy populations vis-à-vis a delimitation of the (racialized) other, were bound up with the perceived role and virtues of the citizen-soldier. Preparing for war became rational and routine. Soldiers, both as individuals and as an imagined category of socially constructed persons, chiefly become to be defined in relation to the stuff of the body: the spilling of blood, origin stories, primordial ties and fraternity, physical and emotional suffering, debt owed by and to the state on behalf of the poor and labouring classes, conscription and allegiance to the flag, collective memorializing, and the potential for death. Soldiers reflect the ultimate giving of the self through the destruction of the body for the productive purposes of the nation and its collective security.

In the modern period, permanent armies became an integral feature of a state's defence planning and thus a part of a wider rational, bureaucratic, and centralizing undertaking, in a context of widespread capitalist development. While states began to take on some of the qualities of "nation-states" in the sixteenth and seventeenth centuries, their territorial and sovereign power did not become fully expressed until the eighteenth century when their military function expanded (Mann, 1996). At the same time, nations became imagined in terms of sacred idioms and the will to "collective sacrifice" was (albeit at times coercively) established (Balakrishnan, 2009). These developments worked to produce a normative relationship between the soldier and the state understood as a form of exchange marked by notions of mutual responsibility, often referred to as a social contract: the citizen-soldier sacrificed their individual self through military service and, potentially permanently, through injury or death, so that ideals such as the public good might flourish. In exchange, the state would provide the institutional services to care for the soldier and their dependents in the aftermath of the soldier's service to the state: his legacy might be memorialized, reifying soldiers' privileged acts of sacrifice[3] for the nation.

In the context of a developing inter-state system and in the aftermath of the Treaty of Westphalia, North American and European states worked to develop a robust military and administrative apparatus to territorialize sovereignty and preserve their spatial expressions of power (Cowen & Gilbert, 2008; Cowen, 2008). The economic and ideological prerequisites of a standing army and a mass draft allowed Western states to continue to centralize their mechanisms of control so that they could properly implement and execute a system of national governance. Key here is how the state could mobilize citizenship as an aspect of nationalism so that it could install effective technologies of government (Isin & Turner, 2002, p. 6).

The blending of citizenship and nationality, according to Isin and Turner (2002),[4] marked a turning point in the citizen-soldier's subjective orientation towards the state in that one could ostensibly articulate and express a newfound national-collective consciousness. This subjective transformation of the self towards humanist enlightenment principles, alongside a nationalized identity that constructed imagined communities, connected formerly disconnected people through acts of military sacrifice. While again this consciousness can be traced back to the French Revolution, the social contract that was produced because of the crystallization and institutionalization of sacrifice had its heyday in the post–Second World War era, exemplified, in part, by the GI Bill. This bill, which authorized the building of hospitals, low-interest mortgage rates, and tuition stipends, marked the beginning of thriving welfare-state policies aimed at soldiers and their dependents. This expanded the idea of the social contract, a relationship based on mutual reciprocity between the state and citizens, consolidating it in the public's imagination.

Later, I argue that this social contract as a transcendental ideal takes sacrifice as an assumed given but tends to ease the central disputes between liberal and republican views of citizenship. Liberal citizenship, for example, involves the idea that private property is the quintessential condition for the realization and protection of individual freedom (Schuck, 2002, p. 133). The citizen is primarily a private individual. Often noted for its use of possessive and methodological individualism, liberalism advocates for a subject position that maximizes individual liberty and minimizes state attempts at curtailing those liberties.[5] Liberalism promotes a deeply ingrained, possibly irreconcilable opposition between individual freedom and the state: The state must always be kept in check because it is viewed as a set of practices within a given bureaucratic apparatus that is always-already keen on undermining the rights, freedom and liberties of individuals. Simultaneously, the state relies on the coercive-ideological and practical incorporation of the figure of the night watchman, allowing it access to legitimate violence, from surveillance technology to outright military force, to preserve this protection myth.

Liberal constitutionalism, supported by notions of individual responsibility and the institutionalization of the rule of law, attempts to curb the dangerous power of both private groups and state bureaucrats so they do not infringe on so-called public values and public institutions that underwrite the ideals of individual liberty. Following Schuck's (2002) invocation of Locke, liberalism's ideal subject, a person with reason, can depart the state of nature, obtaining freedom through reason. Liberal citizenship considers private property to be the quintessential condition for realizing and protecting individual freedom (Schuck, 2002, p. 133), where citizens relate to each other as embodiments of private property. Citizens freely enter the public sphere to cultivate economic contracts with one another mediated by uninhibited and open markets. This notion of an individual consumed by the need for privacy and economic betterment prevails in dominant liberal thinking and institutions. These public contracts conceived among private people foster a regime that champions the acquisition and protection of property, which is thought to be the legal-institutional, symbolic, and material preconditions of liberal freedom.

Civic republicans, in contrast, claim that liberalism's privileging of private interests, materialism, and normative neutrality reflects its failure to garner a necessary ethical, active orientation to politics that will ultimately sustain the republic and the common good (Schuck, 2002). In their view, liberalism presents an impoverished version of citizenship. Republicans also tend to argue that politics must be conducted and struggled over in a visible public sphere to stave off the state's inherent tendencies towards corruption. This tradition differs from the liberal approach because it emphasizes that individuals should submit to the demands of the public sphere. A public-spiritedness constitutes a republican's orientation to politics, but this spirit is only born, à la Tocqueville, out of habitual practice. Dagger (2002) defines this as characteristic of the integrative and educative aspects of republican citizenship, which folds the individual citizen's actions into a common expression of the good and thus re-inscribes these sociopolitical values and ideals onto citizens and potential citizens alike.

Liberal and republican philosophy alone cannot explain the breakdown of citizen-soldiering and the sacrificial cults integral to the nation-state's claim to legitimate violence. Instead, following Aihwa Ong (1999), I understand liberalism not as "a political philosophy but an art of government … not something that can be reduced to a perfect realization of a doctrine called liberalism; rather, it includes an array of rationalities whereby a liberal government attempts to resolve problems of how to govern society as a whole" (p. 195). As I claim in the following section, if sacrifice is to be understood as a problem of liberal governing, then both unmodified liberal and republican traditions will have little explanatory power. For now, it can be said that, in shoring up the power of war and security discourses, states use violence to justify their authority. War and security, which today I argue are processes that are increasingly fragmented, individualized, and privatized, are thus enabled by neoliberal citizenship regimes.

Preceding the neoliberal period, the welfare-state-inspired social contract indicated a spatial order that was considered national and thus it engaged social relations within the boundaries of the state. For more illustration, let us take an additional historical step back to say that the nation-state system was born in tandem with the citizen-soldier archetype and its associated expressions of symbolic nationalized performance, which hinged on a complex imaginary of sacrificial discourse. In this vein, McClintock (1993) argues that nationalism is transmitted primarily through fetishistic spectacle. Nationalism's translation occurs in the ceaseless circulation of fetish objects, such as "flags, uniforms, airplane logos, maps, anthems, national flowers, national cuisines and architectures, as well as through the organization of collective fetish spectacle- in team sports, military displays, mass rallies, [and] the myriad forms of popular culture" (p. 71). The military and its citizen-soldiers are actively involved in McClintock's discussion of the reification of fetish objects. Likewise, the citizen-soldier as a model is deeply rooted in a political imaginary that renders nation-states as discrete, territorially contained entities opposed to other discrete states. National armies facilitated the symbolic and material development of the nation-state and were a product of a newfound securitized technocratic reason. This rationality required that citizen-soldiers undergo systematic physical and moral discipline so that, ideally, their own vitality would solidify the overall strength of the military and the population.

These multiple ideological imperatives mobilized hegemonic narratives of soldiering and warfare, which included the disciplining of "good" citizens. These imperatives extended beyond the military to schools and factories as per the expansion of capitalist ideals and spaces of production. For example, military work was central in shaping the principles of Taylorism to revolutionize industrial production, workplace organization, and worker discipline (Cowen, 2008). Military logic has fluid boundaries, producing similar rationales and techniques of power across social, political, and economic domains.

In the post–Second World War period, citizenship discourse became a public concern in North America (Brodie, 2008), where the link between citizenship and soldiering was tightly woven together. The public called on the state to provide public education, an unprecedented social intervention justified by the idea of children as a pool of future citizens who would contribute to democracy and nation building and who were thus targets of moral regulation appropriate to potential future soldiers and wives of soldiers (Brodie, 2008). The diverse social movements of the 1960s in North America and Western Europe witnessed the demilitarization of these benefits, and as such, they were extended to increasingly more civilians. In 1966, social citizenship in Canada, for example, was reflected in a threefold policy approach: the Medicare Act, the Canada Pension Plan, and the Canada Assistance Plan. Yet the concept of national duty informed the discourse on social rights. A rights-based discourse[6] framed the parameters of the social contract as an exchange between state and citizen – a reciprocal give-and-take of national duty in exchange for social entitlements (Cowen, 2008). In other words, the state has special claims on its citizens (claims to loyalty and potentially to military service), while the citizenry has special claims on the state (rights of entry and residence, rights to political participation, social, economic, or cultural rights, or claims to diplomatic protection abroad; Brubaker, 1992, p. 63). This exchange also normalized the male-breadwinner family model through the notion of a family wage, thus constructing the "real" citizen as an autonomous masculine subject.

An analysis of the citizen-soldier as a working and labouring body, a body that required care and maintenance, was made possible in this context, and as such, the citizen-soldier was the first body to receive these benefits. The citizen-soldier embodied the highest expression of sacrifice, and thus citizenship, and so served to signify proper conduct for civilians (D. Burchell, 2002). Military sacrifice, as it unfolded within the parameters of a strong welfare state, therefore included an idea of mutual reciprocity. As Cowen (2008) explains, the citizen-soldier provides a peculiar challenge to the concept of democracy, specifically because an explicit denial of democracy marks the citizen-soldier's sacrifice. Soldiers, due to their reliance on a deeply ingrained military hierarchy, and thus a denial of individual rights, are socially constructed *in the absence of democracy.* Ironically, while the absence of democracy marks citizen-soldiers, they are also folded into national narratives about the unquestionable desire for democracy, further legitimizing support for foreign wars abroad. In asking "what is a people and how is it made" (Hardt & Negri, 2000, p. 102), we find that "the modern conception of the people is in fact a product of the nation-state and survives only within its specific ideological context" (p. 102). Accordingly, nation-states developed with an interest in securing territory, acquiring resources to secure war-making abilities, regulating populations, and categorizing individuals based on ethnic markers and nationhood. Sacrificial idioms, as tied to the archetype of the citizen-soldier, were critical to implementing and justifying these processes.

As I argue in this book, military privatization and drones are together symptomatic of the weakening link between citizen-soldiering and sacrifice, thus revealing the decay of this often taken-for-granted figure in nationalist imagory. The significance of the abandonment of the citizen-soldier model of warfare should not be underestimated. Without a robust connection between soldiers and sacrifice, our traditional framework for understanding the social and political meaning of war is eroded. In the following, I discuss two challenges to the figure of the citizen-soldier that have emerged in recent years. Flexibility, or flexible citizenship, is one. This shows one pathway whereby citizenship has fractured ties with soldiering. The second is the figure of the suicide bomber or martyr, which directly challenges the liberal way of war.

Challenges to Soldier-Citizenship: Flexibility and the Martyr

Recall that, in the democratic republican tradition, sacrifice has a reasonably expected place in the highly publicized imagining of social relations and therefore informs the construction of "good" citizens. The will to sacrifice is to be expected as a chief quality of not just soldiers eschewing the danger of liberalism's preference for privacy by acting on behalf of the public good but also their families and ordinary people, too. This section argues that both liberal and republican traditions fail to conceptualize the problematic of soldier-citizenship and sacrifice in the global era, as revealed by the following two related empirical phenomena. The first is the globalization of citizenship, here discussed as flexible citizenship. The second is the figure of the martyr in the context of the global war on terror. To examine the status of the citizen-soldier model of warfare, I next turn to Ong's work on flexible citizenship, followed by Talal Asad's work on suicide bombing.

My argument in this section is that both flexible citizenship and the soldier/martyr archetype provide direct challenges to the meaning and stability of the "good" citizen-soldier as both an essential or undisputed figure in the national imagining of war and a body with assumed immanent ties to sacrificial idioms and cults. While flexible citizenship undermines the public spiritedness championed by republican philosophy, the construct of the martyr as discussed by Asad (2007) tests the ostensible ethical superiority – through its invocation of secular humanism – of liberal violence. Suicide bombing – "the martyr" – is a direct assault on the routinization of privacy and ordinariness of legally justified killing abroad as advocated by liberal constitutionalism. To this end, we ought not to ignore how, "today, cruelty is an indispensable technique for maintaining a particular kind of international order in which the lives of some peoples are less valuable than the lives of others and therefore their deaths less disturbing" (p. 94). Consider the "unbearable intimacy shared in their final moments by the suicide bomber and her or his victims. Suicide bombing is an act of passionate identification- you take the enemy with you in a deadly embrace" (p. 66).

But first, let us consider the meaning of flexible citizenship and how it serves as a challenge to the applicability of the archetype of the citizen-soldier. Popularized by Ong (1999), the term *flexible citizenship* refers to the

> cultural logics of capitalist accumulation, travel, and displacement that induce subjects to respond fluidly and opportunistically to changing political-economic conditions. These logics and practices are produced within particular structures of meaning about family, gender, nationality, class mobility, and social power. (p. 6)

Flexible citizenship shows how nation-states competing in a global economy might seek to separate the identity of the "citizen" from the identity of the "soldier," leading to a new regard for the labour of soldiers as that which ought to be treated like any ordinary commodity circulating in the free market. This labour can be stripped from the body of the citizen-soldier, thereby eliminating or curtailing whatever "rights" that soldier might claim. The effects of flexible citizenship can also be seen from the bottom up, particularly in how former military personnel take up jobs in the private sector, in countries other than their home countries, to gain more stable and lucrative employment.

Ong argues that capitalism compels people to regard themselves through identity categories. There emerges a tension between flexible citizenship (which requires elasticity) and identity (which tries to fix the self against different categories of being). Ong invokes the concept of transnationality to combat a tendency in social thought to define the "global" as political and economic and the "local" as cultural. The concept of transnationality allows Ong (1999, p. 4) to capture the horizontal and relational qualities of embeddedness within socio-economic and cultural processes. Flexible citizenship requires people to embrace a nomadic existence and be willing to enter and exit labour markets on a whim and restructure their relationships so that they might maximize economic productivity within and beyond multiple national borders. Combining the insights of Marx and Foucault, Ong asserts that the art of government has been responsive to the challenges of transnationality, thereby maintaining the nation-state as a powerful force in the globalization of capitalism. Hence, Ong does not sympathize with a post-nationalist order. Ong asserts that Eastern states, for example, have contradicted the "hegemonic link between whiteness and global capitalism" (p. 181) by adopting practices of "graduated sovereignty" and by making use of "differentiated zones of sovereignty" and "sovereignty free zones," as exemplified by growth triangles in Asia. Subjects are thus unevenly included and excluded under different forms of sovereignty so that flexible production through regional reorganization of economic activities can occur. By breaking the dichotomy of "traditional" and "modern" economies, Ong shows how Asian countries differently assimilate capitalism, dispelling the popular scholarly notion of core, periphery, and semi-periphery nations.

While her empirical interest is focused on Asian economies, her theoretical insight into the concept of flexibility is nevertheless applicable here. Using flexible citizenship as a challenge to the archetype of the citizen-soldier, particularly as it is conceptualized within the republican tradition, one main point can be asserted: The logic of flexible citizenship has disrupted what republicans would regard as a sacred bond, that between the citizen and the state, a bond most acutely tested in the will to sacrifice as seen in the publicized theatre of war. Although the violence of war continues, as I show, the outsourced, disembodied, privatized and otherwise invisible qualities of this violence, as it is executed within the parameters of neoliberal governing, mean that the model of the citizen-soldier endures an unbearable strain due to the increasing distance between the subjects and objects of violence. "Flexibility" has pushed many of the qualities of military violence into the private sphere. Wars of the globalization era remind us that, in this age of global mobility,

> military operations and the exercise of the right to kill are no longer the sole monopoly of states, and the "regular army" is no longer the unique modality of carrying out these functions. The claim to ultimate or final authority in a particular political space is not easily made. Instead, a patchwork of overlapping and incomplete rights to rule emerges, inextricably superimposed and tangled, in which different de facto juridical instances are geographically interwoven and plural allegiances, asymmetrical suzerainties, and enclaves abound. (Mbembé, 2003, p. 31)

Military privatization is both an effect and a contributor to this patchwork of rules. Here, the relationship between citizenship and sacrifice is weakened in light of the military's use of flexible citizenship. This captures a shift relative to the sacrificial logic bound up with the archetype of the citizen-soldier, witnessed by the bureaucratization and incorporation of experts into the machinery of government. Private military corporations, as this book shows, trade on expert knowledge. This expertise requires a parallel shift towards flexible citizenship, which can complement and accommodate the terms of neoliberal governmentality. The neoliberal demand for both flexible citizenship and free-market capitalism (synthesized in the act of privatization) has pierced the military sphere, where private contractors resolve the problems of personnel shortages required by netcentric warfare and related deficiency in expertise. Perhaps most importantly for our purposes, private military contractors are not subject to the benefits normally afforded to military personnel as set out by the Department of Veterans Affairs.

Military outsourcing reshapes the very need for sacrifice in a period marked by flexibility and de-nationalization: States can now surpass borders and harness the appropriate forms of technology to justifiably incorporate non-humans and non-citizens into the practice of warfare vis-à-vis the interrelated logic of free-market capitalism, the decline of the mass army model, and flexible citizenship, promoting various forms of governing at a distance. Two recent examples include Russia inserting ten thousand North Korean troops into its war machine; meanwhile, the Australian Defence Force has recently opened its ranks to non-citizens. Rather than the democratic republican sentiment of a shared fate energized by the collective bloodshed of patriots in the name of the nation (a public spiritedness), we see how flexible citizenship complements neoliberal governing to alter how the state perceives soldiers as politically costly, and inherently cognitively and physically vulnerable, but yet sometimes necessary liabilities. Moreover, we contend with how soldiers, and ordinary citizens too, conceptualize the meaning of their potential for state violence, which was traditionally justified a priori by their very citizenship status. This is what is at stake in the decline of the citizen-soldier model of warfare. Right or wrong, nations wage war with reference to the citizen-soldier's actions as reflecting the highest expression of sacrifice in defence of *everyone*.

For Asad, the citizen-soldier is an important figure in his 2007 text *On Suicide Bombing*. Asad, writing against "liberal moralists," motivates the reader to theorize how we might more thoroughly comprehend the liberal West's culture of war through the "civilized" citizen-soldier, in relation to the figure of the "uncivilized/barbaric" suicide bomber. Of the citizen-soldier, Asad writes:

> There is another, less dramatic aspect of modern state violence to which I want to draw attention and that informs liberal politics. The mobilization of individuals within and by the sovereign democratic state and the care devoted to its population have been at the heart of the liberal conception of the good life. And a guarantee of that life is the citizen-soldier who is prepared to kill and die for it, yet whose health, longevity, and general physical well-being are *objects of the democratic state's solicitude* [emphasis added].[7] (p. 60)

In contrast to the civilized citizen-soldier is the martyr, a figure who reveals and agitates within secular humanism the tensions

> that hold modern subjectivity together: between individual self-assertion and collective obedience to the law, between reverence for human life and its legitimate destruction, between the promise of immortality through political community and the inexorability of decay and death in individual life. These tensions are necessary to the liberal democratic state, the sovereign representative of a social body, but they threaten to break down completely when a sudden suicide operation takes place publicly and when its politics is seen not to spell redemption but mutual disaster. (Asad, 2007, p. 91)

What sets these two figures apart is their relation to sovereign power, or that "unique act of freedom" (Asad, 2007, p. 67). The citizen-soldier is, in Asad's telling, the product of the nation-state's love and of the care and devotion that has historically marked how the nation-state has cultivated and mobilized him to protect the good life. To protect life, liberal secular humanism is informed by the notion of redemption, which asserts that some humans must be treated violently so that humanity might be redeemed. "It is a part of the genealogy of modern liberalism itself, in which violence and tenderness go together" (Asad, 2007, p. 88). In this construct, suicide is equated with sin:

> In the Abrahamic religions, suicide is intimately connected with sin because God denies the individual the right to terminate his own earthly identity. In the matter of his/her life, the individual creature has no sovereignty. Suicide is a sin because it is a unique act of freedom, a right that neither the religious authorities nor the nation state allows. Today, the law requires that a prisoner condemned to death be prevented from

> committing suicide to escape execution; it is not death but authorized death that is called for. So, too, all other convicts in prison, all soldiers in battle, and the terminally ill cannot kill themselves, however good they think their reasons for doing so may be. The power over life and death can be held legitimately only by the one God, creator and destroyer, and so by his earthly delegates. But although individuals have no rights to kill themselves, God (and the state) gives them the right to be punished and to atone. (Asad, 2007, p. 67)

Suicide is an act equated with an expression of sovereign power, an aggressively controlled expression of freedom that in the citizen-soldier model of the politics of killing and dying is ultimately reserved for the nation-state and its "earthly representatives." This notion of how self-immolation is wrapped up in a politics of sovereignty is comprehended through the logic of sacrifice. "One must urge the citizen-soldier to give up his life so that a particular way of life may be reproduced – a sacrifice" (Asad, 2007, p. 85). Citizen-soldiers on some level must agree to the possibility of death on the battlefield. Importantly, this is an authorized potential death, containing the signature of the sovereign in the act of sacrifice.

However, in the suicidal, "illegal" actions of the martyr not only is this notion of the nation-state's possession of sovereign power flagrantly contested in the act of self-immolation through surprise bombing, but the final act of the story as told by the Crucifixion[8] – that of redemption and subsequently humanity's potential for moral improvement – is circumvented, thereby *denying the act of violence its sacrificial potential.* In the disruption of the liberal narrative of wartime sacrifice, the martyr's attempt to control the context and outcome of sovereign violence reveals how the figure of the citizen-soldier, that is, its very applicability in contemporary war, hinges on these paradoxes within the liberal West's culture of war. This is specifically apt in its telling of the relationship between the nation-state, sovereign power, and citizens in the management of life and death. When people like Bill Owens and Cindy Sheehan challenge the state in their telling of sacrifice, they are denying the state's right to use sacrificial politics as a vehicle to normalize war's violence in the context of a wider redemption narrative.

Asad (2007) wants to move away from "the mythology of suicide [which] encourages fantasies of accessibility" (p. 41), thereby contradicting those which seek to uncover motivation, and/or frame suicide bombing as a religious sacrifice. For Asad, the martyr's connection with sacrifice is contingent. Suicide bombing is not always-already an escape from political oppression, and/or an effect of a Freudian death wish. Mainstream interpretations, even those thought to be nuanced and scholarly, tend to see suicide bombing as a violent expression of either a perverted totalitarian Islam, or of a primordial religious urge that secularism has overcome (p. 95). Key here is the ostensibly secular, liberal West's insistence on its "humane" means of killing as being not only justified but also superior to the violence expressed by the terrorist and terrorizing actions contained in suicide bombing.

Asad's theoretical framework allows him to examine suicide bombing through the intersections between human finitude, violent death, and the re-establishment of political community. For Arendt (as discussed by Asad, 2007, p. 56) the pursuit of immortality is inextricably linked to a profoundly this-worldly endeavour – the founding or re-creating of a just community on earth, and as such, "political life is the space of an earthly permanence that can compensate for human mortality" (p. 58). But Asad's (2007) inquiry is less concerned with moralistic narrativizations of suicide bombing than with how "violence is embedded in the very concept of liberty that lies at the heart of the liberal doctrine" (p. 59). An analysis of the link between personal mortality and political action reveals how the citizen-soldier and martyr, rather than purely discrete and oppositional categories, in fact share tenuous interactions with iterations of sovereign power. To exercise sovereign power is to control mortality (Mbembé, 2003).

It is sovereignty's link to sacrifice that unexpectedly connects each figure in the imagination and practice of re-founding collective violence. Typically understood as diametrically opposed in terms of what bodies are regarded as capable of wielding legitimate violence, Asad exposes the different understandings of sovereign and biopolitical power that become attached to violence qualified as either legitimate or terroristic. The suicide bomber, in eschewing the typical soldier's uniform and displaying no weapon, brings the destruction of violent death into the ordinary spaces of everyday life (Mbembé, 2003, p. 36). Alternatively, the citizen-soldier, whose cost of survival is calculated in terms of the capacity and readiness to kill someone else, namely, to impose death on others *while preserving one's own life,* reflects a logic of heroism as classically understood: to execute others while holding on to one's own death at a distance thus consolidating the moment of power and the moment of survival (Elias Canetti, cited in Mbembé, 2003, p. 37). Suicide bombing, in contrast to the citizen-soldier, is therefore not concerned with heroism but rather with challenging the narrative that binds sovereignty to mortality. This discussion has shown how the "romance of sovereignty ... rests on the belief that the subject is the true master and controlling author of his or her meaning" (Mbembé,

2003, p. 13). This romance fixes legitimacy and control over life and death within the boundaries of the nation-state writ large and particularly the body of citizen-soldier.

Summary

The citizen-soldier emerged as a critical ideological figure in the context of nation-states' relation to war in North America and Western Europe. In this period, the model of citizen-solider warfare was anchored in the territorialization of sovereignty, which was bound up with the exteriorization of war. The citizen-soldier, as someone whose sacrifice is thought to de facto constitute national service, served dynamic socio-political purposes for the nation. It did so in a period of state building when nationalism and citizenship were discursively woven together, through origin stories, racialized conceptions of otherness, and territorial, imperialist expansion. As an archetype of citizenship, the citizen-soldier normalized a belief that soldiers' actions in wartime reflected the highest echelon of sacrifice. Given the powerful historical legacy of the French Revolution, military sacrifice is now an assumed given in war. Problematically, the social, political, and historical particularities of military sacrifice, and the archetype of the citizen-soldier therein, are rarely scrutinized or challenged.

This chapter introduced a core point of this book: the idea that military sacrifice ought to be challenged conceptually. It explored whether military sacrifice is essential to war, asking what is at stake if we treat sacrifice as historically specific and thus not guaranteed in war. This chapter sketched the conceptual landscape on which this book rests. I introduced military sacrifice as it relates to drone warfare, challenging its universality to ask what the effects of a post-sacrificial imagining of war might be. The history of wars in the twentieth century shows a turn away from "heroic" and "noble" sacrifice (associated with legitimate wars) towards a post-Vietnam concern with casualty aversion, which included a biopolitics aimed at protecting troops in an era of limited wars/wars of choice. This has troubled the meaning and place of military sacrifice, because sacrifice resonates most with sovereign power (even as it survives under biopolitics).

As this book shows, military privatization and drones – now prolific and increasingly viewed as not only uncontroversial, but obligatory – emerged as two sides of the same coin: the latter being a technological representation of the political philosophy that underwrites the former. Put differently, the contemporary legacy of the drone is to be found in the politics of military privatization. Both practices together help mitigate the political problem of conventional troop deployments – to disavow sacrifice. But in doing so, sacrifice, which comes to be through sovereign power and its embodied, violent encounters, becomes eclipsed. This is a problem because of how central sacrifice is to our theory and practice of war in a liberal democratic society, even if often made implicit. Can we have war without sacrifice? Should we aspire to this? If we extend the logic of drone warfare to its extreme under the banner of humanitarian war, then the answer could be *yes*. This book ultimately suggests that this would present a series of problems. To help illustrate some of these problems, it traces the politicization, governing, displacement, and loss of sacrifice through the prism afforded by the drone and its social and political origins in the philosophy of military privatization.

Overview of Chapters

In **chapter 2**, I argue for a critical political-sociological approach to sacrifice that is concerned with the place and positioning of bodies in the organization of martial violence, how sacrifice is mobilized or immobilized in political and militarized discourses, and the changing meaning of the citizen-soldier archetype. I compare Canada and the United States to explore what I call a paradox of sacrifice to ask how sacrifice becomes depoliticized, tracing under what conditions it includes or rejects bodies or technology. In **chapter 3**, I claim that governing this paradox requires reconciliation between sovereign and biopolitical power. But the governing of sacrifice exploits a confusion between monarchical sovereignty and democratic sovereignty. For military sacrifice to make sense and survive alongside biopolitics, it must on occasion draw on certain aspects of an outdated monarchical sovereignty. To gain credibility, sacrificial discourse today invokes aspects of premodern sovereignty to maintain the nation-state's grip on the sacred via notions of legitimate violence. In **chapter 4**, I trace sacrifice's displacement across notions of publicity, surplus, and embodiment. I show how military privatization is based in a philosophy of neoliberal rationality, which produces two intertwined effects: (a) the rejection of the surplus in war (defined as that which can be harnessed by the nation-state for utilizing war as a site of collective identity) and (b) the desire to perfect and normalize disembodied violence. Together, these effects create the foundation for drone warfare, which finds its origins in the philosophy that underpins and justifies military privatization. In **chapter 5**, I argue that drone warfare destabilizes traditional means of witnessing, knowing, and commemorating wartime violence, and limits how wartime state violence can be recognized in a distinctly sacrificial language. I argue that drone violence gains credibility through a dialectic of visibility (the battlefield) and invisibility (the violence) and renders that which was previously unseen visible, casting light onto the battlefield. However, the political and social landscape in which drone warfare operates and gains credibility is dependent on making *invisible* the scene of violence and the identification of the witness. In undoing the category of the witness, mechanisms to know and critique war are also severed. This is a problem, because knowledge and memory of war are required to critique, stop, and prevent future wars. In **chapter 6**, I conclude that, rather than liberal democracy's promise of a decline in violence, drones and military privatization contribute to the expansion of more, increasingly invisible, and therefore increasingly meaningless violence. The concept of "antisocial war" captures these trends in state violence, where privatization and drones work together to destabilize the archetype of the citizen-soldier and its place in warfare. This is not to say that antisocial war does not harm or involve people, or that it has no social effects. Rather, I deploy this phrasing to suggest that the future of war is to be found with a view to the past. Antisocial war invokes aspects of eighteenth-century war-making practices, even as these practices materialize through emerging and futuristic technologies.

Chapter 2

Politicizing Sacrifice

U.S. forces will no longer be sized to conduct large-scale, prolonged stability operations.
– *Leon Panetta (US Department of Defense, 2012, p. 6)*

Sacrifice deals with humankind.
– *Girard (1979, p. 90)*

Advancing a Political-Sociological Approach to Sacrifice

What does it mean to take seriously the place of bodies in war? This chapter examines political discourse about war, tracing the inclusion or exclusion of sacrificial language and the effects. In contrast to traditional understandings of sacrifice, I argue for a political-sociological approach that can account for power dynamics: How do bodies and technology relate in sacrificial imaginings? I briefly compare Canada and the United States here to illustrate how sacrifice unfolds in political discourse alongside conceptions of the nation-state. Comparing these two countries is particularly useful as they each have very distinct and oppositional approaches to militarization and war, with the United States being a military superpower, and Canada preferring the language and practices of peacekeeping. Despite their proximity, they have notably different approaches to the discourse of military sacrifice. In Canada, sacrifice is deployed in the examples given in a celebratory manner, whereas in the United States, it is deployed in a cursory manner, regarded as important but presented as ideally no longer required in future downsized, high-technology wars. What explains these different approaches? Using this tension as a springboard, I claim that sacrifice ought to be thought of as a productive paradox insofar as it serves multiple purposes, all ultimately aimed at legitimizing martial violence. How might we understand military sacrifice as a paradox? In this contemporary era of humanitarian wars, characterized by peace-driven violence that is also meant to empower life, how citizen-soldier bodies are deployed in this humane architecture of lethal violence – "the materiality of sacrifice" – constitutes the grounds of this sacrificial paradox. The paradox of sacrifice unfolds within the broader problem of how states govern the tension between a biopolitical concern with life and sovereignty's concern with death. While these are two distinct directions of power with different aims, targets, and trajectories, they are tenuously reconcilable. It is through sacrifice that this reconciliation is made possible.

Sacrifice is an enduring interdisciplinary concept within the social sciences and humanities. It has long warranted inquiry from a variety of theoretical gazes: anthropologists and sociologists, philosophers and theologians, legal scholars, and feminists across disciplines, many of whom claim that the founding moments of human society can be traced back to some initial experience of sacrifice. Legal scholar Paul Kahn (2008), for example, states that the original political moment is not a contractual one but a sacrificial one.[1] Sacrifice is the subject of interdisciplinary conversation within the sub-fields of ethics, psychology, and religion, between scholars such as Kierkegaard, Heidegger, Blanchot, Bataille, Derrida, Nancy, Nietzsche, Girard, Levinas, Kristeva, Hubert and Mauss, Irigaray, and Žižek. To various degrees, these scholars develop their theories of sacrifice through an engagement with Lacanian and/or Freudian psychoanalysis, Hegelian dialectics, the history of Christianity, and the Bible. *The Question of Sacrifice* (Keenan, 2005, p. 14) is an example of a remarkable text that extends this interdisciplinary conversation. Keenan methodically investigates a rich genealogy of the theories of sacrifice. This genealogy, he argues, invokes, and requires, a parallel investigation into the history of economics, sexism, and Christo-centric evolutionism. I ultimately depart from Keenan's approach, because as a non-historian, my interest in sacrifice is not confined to the past but instead asks about the conditions of possibility that give rise to sacrifice in the present. As a non-theologian, and non-philosopher, I will also not contribute to these metaphysical, psychoanalytic, or religious conversations, which tend to be preoccupied with a search for origins and theoretical closure. Instead, a political-sociological approach is concerned with the power and politics at play that give rise to everyday understandings of sacrifice.

In this chapter, I suggest that these dominant theoretical frameworks do not readily translate to the empirical context this book is situated in – the post-9/11 era, the war on terror, and its aftermath. These theories cannot capture the nuances of globalized militaristic violence characteristic of the war on terror. Nor can they fully account for the sociopolitical and economic conditions that fix meaning to who can order a sacrifice, who can sacrifice, and who must be sacrificed. Instead of inquiring, for example, about the unstable role of citizen-soldier archetype in relation to "public" and "private" modes of violence, such specificities are omitted in the grand project of constructing abstract universal theories of sacrifice, thereby obfuscating sacrifice's materiality.

Part of the problem with the dominant literature on sacrifice is that it is difficult to imagine a straightforward way, assuming that this is even desirable, to import the deeply religious, meta-historical, philosophical, and psychoanalytically motivated contributions to the concept of sacrifice into a distinctly political-sociological framing of global militarized violence, where biopower (to "make live") is taken seriously as an element of sacrifice alongside sovereign power (to "let live"). We should be able to explain the applicability of sacrifice in the reconfiguration of capitalism's link to war, discussed here as the neoliberal restructuring of capitalism, particularly revealed in relation to military/security privatization, and what I claim is its technological extension: the proliferation of weaponized and/or surveillance drones. In many approaches to sacrifice, fixing universal meaning in the makings of a general theory, rather than viewing sacrifice as contingent, is commonplace. To view sacrifice with scepticism, as that which ought to be rendered in ambivalent rather than certain vocabularies, is to destabilize it. In this act of destabilization, a distinctly political-sociological critique of sacrifice emerges.

A political-sociological approach to militarized sacrifice thus views sacrifice in discursive terms alongside the changing meaning of bodies in war. It focuses on (a) the place and positioning of bodies in an increasingly disembodied architecture of global violence, (b) how sacrifice is mobilized or immobilized as it is invoked or silenced in political and militarized discourses, and (c) the changing meaning of the status of the citizen-soldier archetype. This approach views sacrifice, especially the taken-for-granted moments when it is both implicitly and explicitly invoked politically, as a gateway into the practices and processes of power integral to disembodied techno-fetishistic war, and the various expressions of privatization and flexible citizenship that enable it. My approach to sacrifice is characterized by this broader inquiry into how it, as a fundamentally embodied phenomenon, is constructed and represented in increasingly disembodied forms of violence. A political-sociological approach further explores sacrifice in terms of its discursive qualities; how it fails or succeeds as a political, rhetorical device within given militarized, specialist, and/or institutionalized speech acts. This focuses on how political elites and ordinary citizens alike harness or reject the symbolic capital of sacrifice to more efficiently govern or resist these contentious, exclusionary, and antagonistic relations that constitute the politics of the unequal power relations which frame what counts as sacrifice. Finally, a political sociology of sacrifice mobilizes sacrifice to contribute to discussions on the unstable meaning of the violence associated with the archetype of the citizen-soldier. It reveals how military sacrifice is tied to the image of the citizen-soldier, a historically contingent archetype, which is further tied to the dominant traditions of democratic republicanism. Herein I argue that the link between citizen soldiering and sacrifice often consolidates *retroactively* in reference to nationalized affective frameworks, which reify or destabilize the citizen-soldier as embodying sovereign will as the executor of legitimate violence.

This chapter does not advance an objective theory of sacrifice but rather claims that sacrifice acts as a portal into a larger problematic of how Canadian and American state officials differentially construct, mobilize, or reject sacrificial themes and discourse to then authorize, make credible, and justify distinctive styles of military violence. Thus, this chapter contains a comparative element,[2] but the comparison is only invoked briefly to reveal the malleability and incongruity between sacrifice and the citizen-soldier archetype as it unfolds in relation to notions of history, the sovereign nation, and war. However, it is worth noting that the idea of a just war is integral to the imagining and practice of sacrifice. The drive to sacrifice, in the form of an individual action of self-sacrifice, such as signing up for the military, or as a top-down, nationalized political-economic policy, such as when implementing a mass draft and cultivating a nationalized ethos of widespread sacrifice, is especially apparent when wars are widely regarded as just. Given the ambiguity inherent to the war on terror and the related erosion of temporal and spatial boundaries of counterterror operations, the status of sacrifice, in both its micro and macro imaginings and practices, appears vague. Should a sufficiently moral and just war present itself,[3] sacrifice could re-emerge as a credible and powerful principle.

Defining Military Sacrifice

Military sacrifice can be thought of as the moment where violence and embodiment are joined in a risky encounter with an Other. The resultant injuries or deaths of soldiers can be both destructive and/or productive (Baggiarini, 2019). It is destructive insofar as sacrifice needs bodies. Sacrifice must take, and it must *consume* something, such as bodies and/or limbs, to then prove and validate the sovereign's existence in the public domain through representation and recognition of the sovereign's presence. This recognition then reconfirms the existence of the author of violence. Moreover, sacrificial violence is productive in that it brings into being, through inclusionary/exclusionary practices, those who receive the benefit of the sacrifice, such as citizens and the nation-state. Military sacrifice historically produced a surplus that is productively harnessed by the nation-state for utilizing war as a site of collective identity. Sacrifice is paradoxical because its deployment is contingent on the politics of, and configurations of, violence in which it is embedded. However, as will become clear further on in the book, the dominance of humanitarian-inspired and so-called bloodless wars, the latter of which were especially made credible during the first Gulf War and in the post-9/11 period exemplified by the ideologies of high-technology and casualty aversion, means that sacrifice is being stripped from the body of the citizen-soldier, edged out beyond the terrain of the social, paving the way for forms of non-sacrificial and therefore increasingly meaningless forms of war. Sacrifice then is an embodied, surplus-producing, and publicity-driven practice. Sacrifice offers a portal into the political, and when configured as a paradox, reveals a style of violence that must negotiate its terms via the political quality of life (Agamben, 1998). We encounter today a style of violence that, as it becomes dislodged from the archetype of the citizen-soldier, risks being made increasingly meaningless. This tendency is amplified by theories that overlook power and the materiality of embodied politics and instead mystify sacrifice.

For example, Kahn (2008) argues that the sacred is "self-evident because there is no test of its truth apart from the character of the experience itself" and that sacrifice "never makes sense within the terms of our ordinary lives- that is the whole point" (p. 117). Moreover, Kahn (2011) claims that "the act of sacrifice can refer to the self or an other. Sacrifice is both transitive and intransitive. Whom do we sacrifice for the state: citizen or enemy" (p. 94)? These assertions rightly point to the difficulty in capturing the empirical content of sacrifice. Sacrifice appears to be somehow everywhere and nowhere,[4] yet "the sovereign dies when citizens are no longer willing to take up its presence in their own bodies" (p. 176) and "a state that no longer thinks of itself as capable of demanding a life has passed out of the political ethos of modernity. The narrative of such a state can no longer invoke the myth of sacrifice or the experience of revelation" (pp. 122–123). Khan is correct to draw our attention to the impacts of a state that cannot include citizens in a national politics of sacrifice. But this shift does not indicate the "death of the sovereign" but, rather, its reconfiguration. Khan's analysis risks a deterministic view of sacrificial politics, because it is based on an overstated distinction between law and sovereignty and the place of theology and religious imagery and the meaning therein.

In a similar mystifying vein, Girard (1979) argues that sacrifice creates the necessary conditions for the socially acceptable use of violence. Girard claims that all violence emerges from mimesis, or mimetic desire. Desire is a by-product of imitation and is inherently mimetic. "Man" does not actually know what he wants until he sees that desire in an Other – and this produces rivalry or conflict. If two people desire the same object, for example, they will eventually lose sight of the object and focus their hostilities on each other in the form of violent competition. Continuing in Girard's narrative, eventually everyone becomes a model/rival of the other, increasing mimetic desire and conflict, resulting in a snowballing of violent contagion. When mimetic violence becomes too intense it will be targeted towards a surrogate victim – the scapegoat. For Girard, sacrifice allows for the circulatory movement or displacement of vengeance onto a scapegoat. Unchecked violence is only tamed through sacrifice or through the implementation of a judicial system.

According to Girard, sacrifice is an act of violence without the risk of vengeance. Girard argues that there is a clear identifiable essence that links and homogenizes all sacrifices (granting it a meta-historical character)[5] as far as they all relate back to this mimetic theory of violence such that "all prohibitions and rituals can be related to mimetic violence" (Brighi & Cerella, 2015, p. 12). Furthermore, "where critical theory employs the method of relentless critique and deconstruction to navigate the complexity [of] the late modern condition, mimetic theory becomes decidedly metaphysic and eschatological when it invokes the possibility of grace and redemption via the path of *imitatio Christi.* The narrative that Girard weaves through history is clearly incompatible with the incredulity towards foundations professed by theorists of a critical bent" (Brighi & Cerella, 2015, p. 17) and as such appears "too totalizing and mechanical. It is not sufficiently responsive to the meanings realized in and through the violence itself" (Khan, 2008, p. 120). Thus, we might conclude that Girard's approach is ahistorical and deterministic. Girard's arguments about how violence unfolds preclude and overlook the possibility of that very struggle over interpretation. For Girard, sacrifice is the site whereby violence can be enacted in a healthy way and is thus externalized and purged from the community, thereby protecting the community from its own capacity to commit violence against itself. The purpose of sacrifice is that it serves as a mechanism by which to restore and repair social harmony within the community through a disavowal of outsiders, via the externalization of violence through the scapegoating mechanism as seen in the identification of a surrogate victim.

Yet, in the context of state soldiers, their sacrifices do not produce exclusionary practices but instead the opposite – they produce a community. Their sacrifices are positively remembered and/or lovingly celebrated. The deaths of citizen-soldiers are consecrated within the boundaries of the nation and memorialized in a way that allows for both the production of a shared collective identity as well as a projected future-oriented discourse of unification, empowered through a shared national destiny. In this way, the citizen-soldier is not a scapegoat but a source of sustained community. The death of a soldier does not result in their removal from the community but, instead, solidifies their place within it.

Girard, whose theorization cannot account for the biopolitical or life-enhancing qualities of sacrifice cited earlier, is rightly critiqued for his Hobbesian essentialization of male violence, the denial of possible cooperation, and therefore avoidance of violence among people, making the analytic point of departure and point of arrival foreordained (Taussig-Rubbo, 2011, p. 141). Furthermore, for a political sociology of sacrifice, a straightforward application of Girard's analysis risks romanticizing the extent to which global wars are necessarily contingent on a violent schema of sacrifice, thereby de-historicizing the role that bodies play in the meaning of the sacrificial violence. If, as Girard remarks, "sacrifice deals with humankind" (Girard, 1979, p. 90), then we must ask how the symbiotic relationship between military/security privatization and drones reshapes the imagining and applicability of sacrificial violence considering the changing notions of what it means to be human-in-war. While inspiring the question, Girard's analytical tools alone do not aid in the investigation. Moreover, from Khan and Girard's earlier claims, we cannot glean how sacrifice, as an effect of contentious (but not predetermined) social relations, produces an increasingly uneven application and experience of military sacrifice. Military sacrifice implies an inherently unequal relation between the state and soldiers, and soldiers and civilians: benefits for some and disadvantages for others. Still, much of the dominant literature on sacrifice does not extend to the political context of war or militarization but, instead, limits its focus to a romanticized past: ancient rituals or deeply religious aspects of sacrifice.

Sacrifice is often theorized as primarily a religious act that mediates the domains of the immanent and the transcendent. It is conceived of as an act that allowed a sacrificial victim to bring someone in closer proximity to God because man and God are not in direct contact. This explains how the fusion of the sacred and the profane, which come together in the various elements of the sacrifice, is completed in the victim (Hubert & Mauss, 1964/1981, pp. 11–32). Hubert and Mauss (1964) further argue that "sacrifice is a religious act which, through the consecration of a victim, modifies the condition of a moral person who accomplishes it or that of certain objects with which he is concerned" (p. 13). We might productively translate this point in line with the citizen-soldier, whose death renders sacred not just their sacrifice but also the nation and its cause. Hubert and Mauss show how sacrifice is not simply tied to the logic of social exchange with God but acts as a means of consecrating.

However, globalization has dissolved and re-inscribed the content of citizens and others, obfuscating the categories "victim" and "scapegoat," rendering consecration contentious. As I argue, drone warfare avoids and conceals, rather than exposes and reveals, the characteristic elements of a sovereign power's place in sacrificial violence. Nevertheless, despite complications, we might agree with Girard and others that sacrifice creates the grounds for the transmission of a transcendental truth. It is profoundly communicative as far as it is meant to engage, and bring into being for its receivers, the gift of knowledge that only a higher power can provide – to allow the sacred and the profane realms a chance at contact. Given that sacrifice is partly about the power to communicate, name, and legitimize certain deaths as having public significance, to recognize certain deaths as loss (and others as forgettable) the current trend towards the privatization of warfare and incorporation of private contractors into a denationalized mode of warfare lends itself to theorizing how this contestation over the domain of sacrifice is particularly relevant today.

I argue that the incorporation of private military contractors allows the state to free itself from the accountability and responsibility typically associated with twentieth-century warfare. Henceforth, governments can potentially wipe their hands clean, so to speak, of any hint of burden. That potential burden (or the government's end of the social contract) is thrown to the wind of the market thus becoming a problem and a solution to be remedied by the market. Alas, this is the cornerstone of privatization, but it also suggests trends towards the desacralization of citizenship. To say "no" to sacrifice through privatization schemes, to bracket and deny it its communicative powers, is to create distance between the ordinary citizen and the state's violence it supposedly applies on its behalf. Drones are the technological expression of this sociopolitical distancing and likewise reflect the silencing of sacrifice's communicative powers.

Privatization is an effect of the neoliberal restructuring of capitalism, whereby state and non-state actors alike have an expressed biopolitical investment in the socially constructed, evolving terrains of life and death. The biopolitical problem of liberal government is reflected by the state's commitment to life as being also directly related to its control (or anxiety over its lack thereof) over the communicative potentiality of death. As the previous chapter highlighted, suicide is thus of particular importance because it operates as a direct challenge to the state's ability to control the terms and rhetorical after-effects of those who can die.[6] Part of the state's strategy of managing these rhetorical after-effects, and thus the potential for certain deaths to be contested as sacrifices retrospectively, is the move to privatize sacrifice (Taussig-Rubbo, 2009, p. 86), moving its effects into the domain of the home or private family, thereby neutralizing them by placing sacrifice within a gendered framework outside the public gaze (Taussig-Rubbo, 2009).

Jay (1992) argues that rituals of sacrifice are used to maintain patriarchal relations and reconcile the problem of having been born of woman. In foundational narratives, women are constructed as the symbolic, cultural, and biological reproducers of the nation. Women occupy the home front as the "Good Mother" or "Beautiful Soul," while "the male heterosexual human or citizen is firmly located in the public sphere, dis-associated from the female private sphere, or the realm of necessity of the body" (Lister, 2002, p. 194). A woman's sacrifice is defined relationally or by lack. Her body is in and of itself insufficient, so sacrifice is experienced through men, embodying "men's honor [and] their accomplishments, and are the repository of all that should be protected and conserved" (Pin Fat & Stern, 2005). Sacrificial rituals unite community members by disavowing outsiders. Indeed, the opposition between the sacred and the profane is key to sacrificial rituals: it is often women's and feminized bodies that represent the impure. Naturalizing men's transcendence over the constraints of womanhood, particularly in reproduction, sacrifice is the site at which men's historical need to establish and maintain the integrity of kinship relations is contingent on controlling childbearing women. Jay (1992, p. 37) argues that sacrifice is performative because it causes what it signifies. For Jay, sacrifice gives continuity to the social world, worlds that are often characterized by male domination or explicit misogyny, which is why theories of sacrifice tend to resemble the traditions they claim to illuminate (p. 128).

As such, while they tend to universalize sacrifice these theories also fold into them a universalistic account of gender and patriarchy. In the focus on the gendered, embodied, and performative dimensions of sacrifice, Jay moves us closer to a political-sociological approach to sacrifice. Extending this critical, feminist impulse, Schott (2010) argues that sacrifice reveals the potential of violence that is imminent in society. Schott claims that sacrifice can be used as a diagnostic tool to ascertain how violence informs the political. Still, scholars disagree about whether we can or should attempt to know the internal logic of sacrifice truly and objectively. Perhaps some of these tensions within the literature on sacrifice have to do with the inability to translate what is lost in the theorizing when sovereignty is historicized – put differently – what is gained when sovereignty is de-historicized. What I claim is a mystification and depoliticization of sacrifice on the part of these theories might be resolved with a more nuanced treatment of our contemporary historical moment, considering the specificities enabled by a political-sociological approach.

To summarize so far, the aim of this chapter is to demystify sacrifice, first by bracketing the theological and philosophical arguments, and second by troubling sacrifice by asking: How is sacrifice politicized, problematized, and deployed, and what are the effects? Furthermore, what kind of *work* is sacrifice doing for sovereign violence in the context of biopolitical governmentality? This desire for simplification should not imply a stripping of the complexity of sacrifice's rich conceptual lineage but, instead, reflects an attempt to write about sacrifice from a sociopolitical position – within a specific historical period, that of democratic sovereignty – and to construct a distinctly critical sociology of sacrifice, which ought to be concerned with how sacrificial bodies are included or excluded in acts of military violence. Thus, the position is occupied with visibility and justice and the desire to bring sacrifice's materiality to bear on the broader empirical topic at hand: privatized drone wars. I aim to trace how discourses of sacrificial violence are constructed so that subjects (citizens and others) may be more efficiently governed, giving content to the re-entrenchment of national identity's link to militarization.

Next, I explore examples of the deployment of sacrifice in political discourse, namely official state speeches, and defence strategy documents, which differentially invoke critical themes of technology, history, embodiment, and national identity in Canada and the United States. While the primary focus of this research is the United States, the comparison with Canada is productive insofar as it shows the stark contrast between how the two governments conceptualize the relationship between war, technology, and the body, both historically and for the purposes of future combat operations. I argue that, by politicizing sacrifice, we can better understand the specific qualities of sacrifice in relation to how governments either include or exclude bodies in acts of military violence.

Mobilizing Sacrifice

In the twentieth century, the archetype of the citizen-soldier emerged conterminously with liberal politics – liberal citizenship, economics, and democracy. For example, military work was central in shaping the principles of Taylorism that revolutionized industrial production, workplace organization, and worker discipline (Cowen, 2008). Catherine Lutz (2002, p. 727) argues that the period of twentieth-century industrial warfare not only required the raising of mass armies, but it also centred on manufacturing labour and thus incorporated workers to produce relatively simple guns, tanks, ships, and eventually airplanes.

In constructions of the nation-state, the citizen-soldier represents the supreme expression of sacrifice, and therefore citizenship, which other citizens ought to aspire to. The figure of the citizen-soldier signifies proper conduct for civilians, as people who are seen as sacrificing for a particular body or association are regarded as "true" citizens (D. Burchell, 2002). As Frank Pearce (2010) argues, sacrifice draws the sacred (the realm of God) and the profane (the realm of humans) into direct contact. Typically, these sacred and profane realms are mutually exclusive but through sacrificial actions, men alter their societal position, either through transcendence, exchange, or communication with deities. There are social or spiritual benefits associated with sacrifice. Sacrifice allows nation-states to organize collective violence by using the symbolic excess that resides within sacrifice to assert its authority.

For example, when sacrifice is invoked alongside citizenship, belonging, and wartime remembering, it provides the grounds to effectively recast nationalism in violent terms. There are notable intersections between citizenship (the right to claim a certain kind of life) and sacrifice (the right to claim a certain kind of death). Citizenship includes the practice of claiming rights *to something*: the right to life, the right to a better life, the right to enhance one's life and future potential. As such, it contains an impetus towards biopolitics because of its concern with empowering and enhancing life. While the distribution and access to these rights are uneven, citizenship's effects are widely regarded as positive, while discourses of sacrifice might invoke negative images of suffering, loss, and death. Although sacrifice can have positive, altruistic connotations (e.g., the "willing giving of the self"), this book is concerned with the discursive interplay between both negative and positive connotations. The management of the interplay between citizenship's promise of a better life and sacrifice's call to potentially obliterate life provides the fruit of this chapter. In the following, I draw on examples from political speeches that pivot on this interplay between citizenship, nationhood, and sacrifice.

As a political phenomenon, sacrifice has re-emerged in political discourse with an expression of renewed gravitas attached to it, having been deployed by Canadian and American state officials in speeches to commemorate wars and/or justify novel ways of applying military force. Consider the following Canadian examples, which include three excerpts from three separate speeches. First, on 13 January 2014, the Honourable Shelly Glover, then minister of Canadian heritage and official languages, in a speech directed at veterans and their families, said:

> I look at all of you, and I think not only of your sacrifice, but the sacrifices of those in my family and many of your families ... we wouldn't be the country we are today were it not for their sacrifices and your sacrifices ... the commemorations will recall the sacrifices of those who took part in the two World Wars [allowing] Canadians to better know and appreciate their history. (Canadian Heritage, 2014)

The minister here is drawing a direct link between sacrifice and the integrity of the nation ("we wouldn't be the country we are today"). What kind of country is that? It appears the listener/reader intuitively knows what kind of country Canada is. We might assert descriptors like "reasonable," "tolerant," "honourable," and "peaceful" for deliberating this question. That aside, the minister is suggesting that sacrifice is partly mystical yet has a ripple effect (whose cause/origin is not clear) as sacrifice applies not only to the actions of veterans but also to the actions of their families. Second, and perhaps most critically, through sacrifice, and its implied but unelaborated upon transcendental quality, "we" Canadians come to know national identity through these sacrifices. Sacrifice is represented here not just an incidental aspect of a certain shared history (the history of the Great and Total Wars); it is also much more: It is a gateway into a pure and uncontested historical knowledge of the Canadian Self.

Second, at the same event, Minister of Veterans Affairs Julian Fantino commented that "two generations of Canadians made incredible sacrifices so that we might live in peace and freedom. It is our profound duty to commemorate and honour those who served at home and abroad and made our country the free, democratic country that it is today" (Canadian Heritage, 2014). Like Minister Glover's comments, Minister Fantino implies that sacrifice has an immediate bearing on our shared sense of the political. Specifically, the form of democracy that is familiar to Canadians' understanding of their nationhood is portrayed as the direct result of these wartime sacrifices, eclipsing other notable, although less spectacular examples of social and political struggle, which may have also contributed to the development of democracy.

In other words, liberal democratic nations have a built-in right to call on citizens to sacrifice – and to ensure the democratic health and survival of the nation they are obliged to answer the call. There is no acknowledgment of possible coercion, deceit, or bad blood underpinning this relationship. Peace, freedom, and liberalism, simplistically portrayed as desirable, unproblematic, and interchangeable philosophies, were born from the wartime sacrifices of two generations of Canadians. Thus, it is "our profound duty" to not simply superficially respect these sacrifices but, in fact, also honour, and most importantly *accept*, their sacrifices as integral to the fabric of national identity. As Canadians, to call into question, criticize, or even refuse these sacrifices would be profoundly un-Canadian, and in fact quarrelsome, the political equivalent, for example, of explicitly refusing to don a poppy on Remembrance Day.

Finally, then prime minister Stephen Harper, at an event honouring the 100th anniversary of the start of the First World War, remarked,

> The mud, the blood, and the sacrifices that marked those years left more than a third of these Canadians dead or wounded. Forgive me if I do not dwell on these numbers, the bitter harvesting of suffering and death. We have had a hundred years to contemplate this war. Much has been written on the subject. And yet, what it means to have lived in muck and disease, to fight through mud deep enough to drown a man, to lose thousands of lives in a single day to gain what could be measured in yards … it's also difficult to measure sacrifice. (Harper, 2014)

The proximity of the words "mud," "blood," and "sacrifice" in the opening sentence is striking. Wartime sacrifice necessitates death, injury, and bloodshed; as such, it requires earthly, territorial bodies living in "muck and disease" trained in battle and exposed to the unforgiving and cruel brutality of war and nature ("mud deep enough to drown a man.") But the precise numbers, the specifics in *political* terms of the bodies do not matter in this construction ("forgive me if I do not dwell on these numbers") because, we are to imagine, their suffering and loss are too overwhelming to submit to such crude quantification, especially when considering the gains made, which could only "be measured in yards." Were these spectacular acts of sacrifice – which Prime Minister Harper is simultaneously politicizing and depoliticizing and cannot be truly known or measured – in line with Kahn's notion that sacrifice cannot be known beyond its experience, worth the yards gained? We are not presented with an explicit answer, but the expectation, especially in light of Minister Fantino's remarks about the duty to accept these wartime sacrifices as noble and just, would be an affirmative response. Sacrifice here is painted as banal, because it affected so many people and families, but it is also characterized by its deeply felt yet intangible and unknowable impact on society. Despite having had "one hundred years to contemplate this war" wherein "much has been written on the subject" sacrifice escapes us; we still are nowhere closer to understanding or truly knowing sacrifice, although all Canadians, Harper suggests, much in line with Kahn's (2011) analysis, still feel the pervasively haunting yet unknowable presence of sacrifice.

Taken together, these comments attest to how the discourse of sacrifice, as it pertains to Canadian identity making, although imprecise and mystical, is resuscitated to give militaristic truth, content, and meaning to the current political climate. This climate includes a parallel attempt to use the 100th anniversary of the Great War to re-nationalize Canadian identity and rebrand it as a nation of warriors (discussed later). Sacrifice implies coercive and unequal relations of power, (it is now well known that people of colour and those living in poverty or with precarious citizenship/residency status are more likely to sign up for the military), yet there is no hint of this political reality in official speeches. Despite then prime minister Harper's own admittance that the meaning of sacrifice is ultimately contested (it is "difficult to measure"), he still nevertheless invokes it in concrete terms, consciously drawing on the symbolic power of the image of sacrifice to make universal claims about the essential link between nationalism and militarization, as well as promote a distinct imagining of Canada's historical legacy and Canadian identity as something inherently vulnerable, that is being remade in relation to the historical unbearable weight of wartime sacrifices.

When combined, these excerpts show how sacrifice is made up in a politicized discourse to inspire an *affective* economy, generating varied emotional responses to be sure, but patriotism is clearly a primary desired emotion, which allows elected and unelected state representatives the comfort from which to access the "hearts and minds" of Canadians. The effect is a subtle, disciplinary one: It grants credibility and consent to govern and normalize its perceived monopoly of violence.

We learn that liberal democracy is not only the ideal governing ideology, but it is also the product of many knowable/unknowable ultimate sacrifices. Liberal democracy is imagined here as fragile and deserving of renewed protection through continued sacrifice.

Furthermore, sacrifice is invoked without regard for specificity or precision around its meaning or effects and with seemingly no tangible consequences or challenges to this: to be brief, it is a term that is taken for granted. It is a seductive and all-encompassing word choice, matched by equally powerful and historically charged imagery, deployed to re-nationalize and re-militarize the nation-state's power – to authorize the sovereign claim to violence and naturalize the uneven and contentious relationships between citizens, non-citizens, and the nation-state in a globalized era. Thus, I argue in **chapter 4** that sacrifice ought to be analysed alongside the changing notions of sovereign and bio power within globalized military economies to reveal what the deployment of bodies means for different forms of power. Likewise, moments when sacrifice is alternatively denied, or outright avoided, suggest an opportunity to better understand how state power differentially re-authorizes "legitimate" military violence.

Although Canada currently has a liberal government, headed by Prime Minster Mark Carney, Stephen Harper's conservative government (2006–2015) was long focused on mobilizing and manipulating the concept of sacrifice, and sacrificial rhetorical discourse, to militarize Canadian identity and biopolitics prior to these excerpts from political speeches. For example, Harper summoned the impact of soldiers' sacrifice for biopolitics in a 2006 address to the media. He explained how soldiering is the highest calling of citizenship "not because you are ready to die for your country [but] because you are *ready to live* [emphasis added] for your country" (Harper, quoted in Cowen, 2008, p. 205). This is reminiscent of when, some 150 years earlier in 1863, Lincoln remarked at Gettysburg that soldiers "[give] their lives so that the nation might live" (Lincoln, quoted in Taussig-Rubbo, 2009, p. 88). In 2008, Stephen Harper's "Sacrifice Medal" was introduced to recognize those killed or wounded in hostile action since 2001. The medal marked the beginning of Canada's support of the global war on terror.

The global war on terror produced a series of political, technical, and social effects of control, regulation, and discipline, which are now well documented. Pre-eminent among these effects is the intensification of militarization in Western states. Militarization includes the "multidimensional and diverse set of social, cultural, economic and political processes and practices unified around an intention to gain both elite and popular acceptance for the use of military approaches to social problems and issues" (Rech et al., 2014, p. 48). Militarization draws on violent or techno-scientific mechanisms and ideas to manage a range of problems, from the ostensibly banal remembrance of life, to sophisticated and sweeping military invasions. It implies "the contradictory and tense social processes in which civil society organizes itself for the production of violence" (Geyer quoted in Lutz, 2002, p. 723). Militarization is essentially *contradictory* because our current historical moment is characterized by, as Lutz suggests, the social organization required to produce wide-scale violence. Yet society places tremendous importance on individual human life. Humanitarian wars, for example, are an effect of this rights-based desire to protect life. Following Lutz, militarization occurs at two levels: institutionally and discursively. First, institutions become refashioned to support military pursuits. Second, society's values change in a way to legitimize the use of force (standing armies, possible increase in taxation) and military spending is deeply connected to the development of knowledge in fields such as physics and psychology, thereby redefining ideas about citizenship, sexuality, and masculinity.

Militarization is integral to Canadian identity despite the dominant understanding that Canada is a nation of peacekeepers rather than war makers. As Sandra Whitworth (2004) says of the traditional understanding of Canadian military might, "the Canadian soldier as peacekeeper is not a warrior but a protector. These are representations that fit very well with the more generalized notions of moral purity that pervade Canadian foreign policies towards developing countries" (pp. 90–91). However, the notion of the Canadian soldier, not as a peacekeeper but as a heroic soldier-warrior, became more visible in popular culture during Harper's reign. This was particularly noticeable after 2006, during the height of the violence for Canadian troops serving in Afghanistan (Richler, 2012).

These attempts at militarizing Canadian identity and rebranding Canadian troops as warriors culminated in 2011 when the Harper government launched a campaign to honour the commemoration of the 200th anniversary of the War of 1812. Until 2015, the Canadian government planned to "invest to increase Canadians' awareness of this defining moment in our history" ("Harper Government Launches the Commemoration of the 200th Anniversary of the War of 1812," 2011). According to the government, the War of 1812 was key to ensuring our country's existence and shaping our identities as Canadians today. As "a defining moment in our history," the campaign has a CA$28 million price tag for its events, ($6.5 on television commercials alone) which includes War of 1812 battle re-enactments, special exhibitions at the Canadian War Museum, and collectable coins, among other community, political and business endeavours. (For comparison's sake, consider that Governor Cuomo of New York City vetoed a bill to establish a War of 1812 commission, although he did allocate a much more modest amount, US$450,000, for commemorations). In the campaign's advertising, the War of 1812 is described as an act of American aggression against Canada, even though Canada was not a country until decades after the war. Altogether, the war comprises a fiercely debated series of historical events whose meaning with respect to the campaign has puzzled many people. Consider the transcript from one commercial: "*Two hundred years ago, the United States invaded our territory,*" a narrator says over dark images and ominous music. "*But we defended our land, we stood side by side and won the fight for Canada.*" In the *New York Times*, Professor Andrew Cohen remarked that "the War of 1812 is part of our [American] history, and that's fine. It's turned into a form of propaganda, and it seems to have married the government's interest in the military, with its interest, some would say obsession, with the War of 1812. It's clearly, to me, part of a campaign to politicize history" (Austen, 2012). In the same article, a less critical military historian, David J. Bercuson, is still nevertheless confused, saying, "I'm scratching my head for the last year and asking myself: Why is the government placing so much emphasis on this war?" Even a battle re-enactment scene curator and administrator concluded, "it's kind of weird" (Austen, 2012).

James Moore, then the Canadian minister of heritage, claimed that the purpose of the campaign is to simply encourage Canadians to remember: "Canada was invaded, the invasion was repelled and we endured, but we endured in partnership with the United States. *It's a very compelling story* [emphasis added]" (Austen, 2012). Indeed, the link between the nation, citizenship, and sacrifice is one that needs periodic ideological maintenance through storytelling and deliberate acts of remembrance. When combined, these examples reveal a resurgence of militarized language (curious for a nation universally regarded as a middle power at best). It also reveals the Harper government's attempts to lay sacrificial groundwork through the re-deployment of social memory, arguing for the recognition of the nation-state's fundamental right to demand sacrifice from its citizens.

However, remembrance is a political, contested process. It must include, according to Jonathan Vance (1997, pp. 3–4) disparate yet complementary aspects of fabrication: a piecemeal combination of invention, truths, and half-truths. The result of this assemblage of details is the creation of social memory. Never entirely shared, social memory of war is the product of how "diverse media were used to convey the myth to those people who had not experienced the events themselves and to ensure that a certain version of the war became the intellectual property of all Canadians, not simply those who had lived through 1914–1918" (pp. 3–4). Thus, Vance argues that through nostalgia, or a tendency to look back on the past with a sense of yearning for better times, "the dominant or collective memory of a society is not always (perhaps, even, not often) based on historical fact, but on a set of assumptions of what the past was like" (p. 9), thus propelling Canadians to remember the war "in terms that sometimes bore little resemblance to its actualities" (p. 4). Consider the Vimy Ridge commemoration, celebrating Canada's role in this battle, which took place on 9 April 1917 (e.g., see Baggiarini, 2019). Canadians gathered in France to remember "the fateful battle" and "reflect on its enduring legacy." According to the *Toronto Star* (April 9, 2017), "there was one key similarity between that Easter Monday on April 9, 1917, and the scene 100 years later: Canadians stood together, shoulder to shoulder, proudly and unabashedly as one people" (Austen, 2012). According to Trudeau's speech at the event, "these ordinary and extraordinary men of the British dominion fought for the first time as citizens of one and the same country. Francophones and anglophones. New Canadians. Indigenous peoples. Side by side, united, here in Vimy. The battle was described as a distinctly Canadian effort – "a true demonstration of all the best qualities that Canada represents: individual initiative; esprit de corps; gumption; enthusiasm" (Austin, 2012). At Ottawa's Vimy ceremony, the environment minister said that the event marked "our coming of age as a country" (Thompson & Blanchfield, 2017). One woman remarked that more people of her generation need to remember the heroic sacrifices of those long gone. "The selflessness that they had, the respect they had for each other, and just all the turmoil and the sacrifice that they've done, everything they lost- it was all for this country that they never saw" (Thompson & Blanchfield, 2017). Indeed, these sentiments confirm Vance's point about the synthesis of myth, fact, and fiction, in the making up of Canadian identity pertaining to war. Of course, many of these statements cannot be empirically tested as "true," but nevertheless, the context in which they emerge naturalizes the statements as symbolic capital in the construction of Canada's wartime past and, therefore, future. Memories of sacrifice, and the production of objective facts that inform and pollute these memories, coexist precariously. Both processes are vulnerable to misinformation, ongoing contestation, and re-making.

Before turning next to the American example, it is worth preliminarily noting that there are marked differences and departures in comparison to the Canadian example, particularly regarding how the United States treats the issue of technology, namely, its growing reliance on unmanned systems and combat unmanning and, therefore, its related emphasis on leaner forces.[7] While the Canadian examples emphasize the role of commemoration, regarding wartime sacrifices as sites of, and opportunities for, historical and national memory making in addition to the significance of bodies in battle for interpreting and solidifying Canadian identity, the American examples show a clear desire to circumvent not only traditional means of sanctifying war, for example, Congressional approval, and/or drawing in a national audience but also the physical limitations of the body through expanding technological proficiency. Of course, these examples, and the subsequent analyses of them, should be understood in relation to the vast differences between Canadian and American historical contexts and practices around war-making: Canada, at best, serving as a middle power with a meagre military presence, is often regarded as composed of humanitarian peacekeepers rather than war fighters (despite the Harper government's efforts at rebranding). The United States, in contrast, being born out of revolution rather than loyalty, is universally regarded as the world's unchallenged military superpower. As the single target of the terrorist attacks of 9/11, it is actively, if not unilaterally, engaged in engineering a post-9/11 counterterrorism road map as well as spearheading military and civilian solutions to post-9/11 counterinsurgencies, defined by limits on the size and scope of troop movements.

Consider the evidence. In 2012, the Department of Defense (DoD) released the Defense Strategic Guidance document, outlining its priorities for twenty-first-century defence. Therein, then secretary of defense Leon Panetta describes an anticipated critical shift in defence policy in response to economic austerity and thus within American practices of war-making more broadly. In the introductory paragraph of the *Unmanned Systems Integrated Roadmap FY2011–2036* (US Department of Defense [DoD], 2011), the authors praise unmanned aerial vehicles (UAVs) for their "persistence, versatility, and reduced risk to human life" (p. v) before asserting that the DoD faces a fiscal environment in which acquisitions must be complementary to the DoD's "Efficiencies Initiative," in other words, defence spending must "pursue investments and business practices that drive down the life-cycle costs for unmanned systems. Affordability will be treated as a Key Performance Parameter (KPP), equal to, if not more important than, schedule and technical performance" (p. v). In a strategic guidance document from the same period, the DoD (2012) claims

> this country is at a strategic turning point after a decade of war, and therefore, we are shaping a Joint Force for the future that will be smaller and leaner, but will be agile, flexible, ready, and technologically advanced. It will have cutting edge capabilities, exploiting our technological, joint, and networked advantage. It will be led by the highest quality, battle-tested professionals. (p. 5)

Furthermore, we learn that U.S. forces will no longer be sized to conduct large-scale, prolonged stability operations. As such, the specialized labour contained in the Joint Force is thought to replace the mass quality of the traditional military structure, most importantly shedding the economic burden therein. Instead of citizen-soldiers, who, some military experts caution, are no longer sufficiently equipped to fight,[8] we are presented instead with the squeaky-clean image of "battle-tested professionals" signalling the downsizing, professionalization, and modernization of the military – the impact of which is discussed more thoroughly later in the book.

Critically, Panetta is not declaring that American forces will outright decline to engage in prolonged operations. What he is saying is that they will *not be sized* to engage in prolonged operations. This points directly to the most profound shift in policy: Prolonged operations are expected, as the global war on terror and its architects do not take kindly to spatial, national, temporal, or political boundaries. But these missions will be conceptualized and orchestrated by combat managers invested in efficient low-cost business practices, and designed for special operations technicians, with fewer, less visible troop movements. Elsewhere in the same document, as discussed by Ian Shaw (2013), Panetta stated,

> As we reduce the overall defense budget, we will protect and in some cases increase our investments in special operations forces, new technologies like unmanned systems, space and in particular cyberspace capabilities and in the capacity to quickly mobilize. (p. 7)

That a flexible, agile, and lean military unit, coupled with unmanned systems, is a pillar of the future orientation towards counterinsurgency and counterterrorism is unapologetically clear. Aside from the mildly ambiguous statement about the economic landscape informing the future of defence strategy, which for Panetta, necessitates a reduction in overall budget yet also produces an increase in investment, the signalling of deploying fewer bodies overall, and an explicit desire to increase investment in technology but reduce investments in troops and their mobilization globally, is abundantly evident. The scope of American imperial power reveals that US strategy mainly involves the "globalization of Anglo-American constitutional principles and neoliberal mechanisms of accumulation and economic discipline" (Gill, 2005, p. 24), in addition to its claim to have the power to "decree national and international rules, laws, and norms, whilst reserving 'exceptional powers' for themselves" (p. 24). The assumption is that the United States, from both moral and economic standpoints, has the right to act as a global state.

This global state mentality is manifested as "full spectrum dominance," which is the surveillance and dominance of land, air, sea, and space, and serves as the counterpart to the Bush administration's concept of the new wars of the twenty-first century. Full spectrum dominance entails the militarization of space, information warfare and control over communication nodes and networks. This neo-imperialism is expressed by the expanding empire of military bases, uniformed military personnel, and informants who have been dispersed globally with the aim of policing world order in favour of American economic and cultural interests. Put briefly, this dominance entails "all elements of national power: economic, diplomatic, financial, law enforcement, intelligence, and both overt and covert military operations" (Gill, 2005, p. 35). This totalizing perspective includes the need for not only sovereign power to override the existing rules of war, for example, through pre-emptive strikes against actual or potential enemies, but also policing what the Bush II administration came to call "the arc of instability" (Gill, 2005, p. 35). Under this paradigm, military bases and surveillance technology will surely extend, but the origin of that power will be made opaque: dislocated, fragmented, and relocated.

Pointing to a reluctance to incorporate citizen-soldiers into future missions is not particularly groundbreaking when considered in a vacuum. Yet contemplate it in relation to the introductory remarks of the very same document. President Obama writes, immediately in the opening line, "our nation is at a moment of transition. Thanks to the extraordinary sacrifices of our men and women in uniform, we have responsibly ended the war in Iraq" (DoD, 2012). That the president would so hastily invoke the theme of sacrifice, prior to his operatives detailing a plan, which explicitly advocates for fewer soldiers, combat unmanning, and, therefore, the disavowal or circumvention of sacrifice, is not a coincidence but a rhetorical move to simultaneously distance war powers from citizen-soldiers, while consolidating the power of the executive branch through the invocation of sacrifice. Billing himself as an anti-war candidate, President Obama won the Nobel Peace Prize in 2009. Yet, perhaps unexpectedly, he was a president at war in the two terms he served. According to Landler (2016), Obama tried to reconcile his anti-war position with the reality of governing by "approaching his wars in narrow terms, a chronic but manageable security challenge rather than as an all-consuming national campaign in the tradition of World War II, or to a lesser degree, Vietnam." From this, it appears that, for Obama, the shift from war to security is a form of preventative "health care" for a body politic, which is exposed to a dull and ongoing pain that cannot be resolved by drawing in a national population but instead requires supra-national, discreet, and singular attacks targeting the (oftentimes multiple) sources of national pain. The reluctance to draw in a national population is key. It suggests a breakdown not only in the traditional means of waging war but also the citizen-soldier's role therein.

Returning to the Defense Strategic Guidance, I argue that, in recognizing these citizen-soldiers for their sacrifices, President Obama is, first, falsely equating the end of war to the officially declared withdrawal of troops. Yet, at the time of publication of this document, there were approximately 10,000 US troops remaining in Afghanistan, and 3,500 troops in Iraq (despite the official withdrawal of troops in 2011). Moreover, the number of private contractors working for the State Department grew eight-fold in 2015 (Weisgerber, 2016). What is Obama's evidence that war has "responsibly ended"? Is it an official declaration, the complete removal of all US troops (and US contractors), or is it the complete cessation of violence in any given state? President Obama perhaps is aware of his audience here being primarily composed of military, defence, and technology industry stakeholders and, therefore, does not feel it necessary to spend time unpacking or defending this claim. Regardless, it is also the case that traditional temporal categories – "before" and "after," "begin" and "end" are not even applicable in the global, permanent war against terror. More critically, however, President Obama's comments normalize these sacrifices as expected and dutiful contributions of citizens, in line with his Canadian counterparts who assert the obligation of states to call on citizens to sacrifice and citizens to answer this call. But, unlike the Canadian officials, the more detailed defence strategy outlined further on in the document suggests, at a minimum, a deep hesitation, if not outright discomfort, with the role that citizen-soldier bodies (as sacrificing bodies) will play in future combat operations.

This discomfort can be explained, I argue, in part, by the sacrificial politics of resistance that characterized the early moments of the 2003

Iraq invasion and the failure on the part of the US government to adequately theorize what its desire for bodyless war meant for its ability to control the sacrificial message in light of growing anti-war resentment. Looking at sacrifice directly provides explanatory power here: The concept of sacrifice is analytically meaningful because in death, particularly the deaths of soldiers serving abroad, discourses of sacrifice allow the potential for the authors of sovereign violence, acting in tandem with the credibility of the nation-state, to breathe life back into those who, it claims, perished on its behalf, reinforcing the biopolitical ordering of life and death by propping up narratives that underscore who is authorized to conduct the sacrifice and who must be sacrificed. But this is not a straightforward process. Sacrifice is a critical component of how people are governed in moments when war is being explained or justified, and thus, neoliberal governments are deeply concerned with establishing some sense of monopoly over its discursive trajectories (Taussig-Rubbo, 2009) to control its meaning. This monopoly of sacrifice is challenged by Bill Owens, Cindy Sheehan, and others, combined with contemporary war technologies and practices, and as such reveals the mythical and increasingly weakened underpinnings of military sacrifice. This results in a contestation over the categorization and classification of sacrifice as it pertains to the lives and deaths of citizen-soldiers and military contractors alike.

Much like how Prime Minister Trudeau's government inherited aspects of Prime Minister Harper's policies, it is worth noting, too, that the Obama administration inherited, and indeed extended, aspects of President Bush's national security strategy,[9] namely, its preoccupation with counterterrorism, defined as the "tactical appropriation of guerilla warfare under the doctrine of counterinsurgency developed during the Vietnam War" (Troyer, 2003, p. 262). In response to 9/11, former president Bush was quick to claim that terrorists are heirs to "all murderous ideologies of the twentieth century. By sacrificing [presumably all forms of] human life to serve their radical visions, by abandoning every value but the will to power, they follow in the path of Fascism, Nazism, and totalitarianism" (Bush, quoted in Troyer, 2003, p. 261). The image of sacrifice invoked here by. Bush is used to insult; it is negative insofar as it is regarded as cowardly and barbaric, in contrast to Obama's use of sacrifice, which speaks out in a congratulatory manner. Bush is constructing sacrifice here as an anachronistic form of irrational premodern violence, as occurring outside the national, and therefore, legitimate sacrificial framework. According to him, it is baseless murder, not sacrifice, since its motivations are perceived to be ideological, carried out by illegitimate non-state actors.

The implication of this rhetoric, as discussed by Frank Pearce (2010, p. 52), is a Manichean vision that not only produces the conditions for "othering" but also implies that "God is not only on the side of the United States, but the devil is with its opponents." Zuriek and Hindle (2004) similarly argue that a moral, "demonological" discourse of terrorism emerged after 9/11, alongside the consolidation of sovereign authority against the other branches of government. In this political environment, the emphasis on counterterrorism encouraged the troubling of the legal status of "unlawful" versus "enemy combatants," as well as reconfiguring military tribunals, inviting unilateralism, and the depoliticizing and persecuting political opposition. It also produced a moral discourse, which viewed terrorists[10] are unhinged evildoers, thus rendering them rightful targets of sovereign violence.

The Paradox of Sacrifice

To consider sacrifice as a paradox, we must clarify the context in which this paradox emerges, namely within the crisis of postmodern warfare. Postmodern wars reveal a crisis as they increasingly rely on technology for full-spectrum dominance, and yet the most powerful military technologies cannot be used (Gray, 2003, p. 215). Speed has replaced geopolitics as the basis of military coordination (Der Derian, quoted in Hooper, 2001, p. 111), extending (simulated) battlefields beyond national borders (Uesseler, 2008). This illuminates the amalgamation of public (state) and private (non-state) powers to realize the goals of liberal wars of choice – wars conceived of in humanitarian terms in terms of empowering life yet nevertheless enabling and perpetuating material destruction. Through the earlier brief comparison between Canadian and American use of sacrifice in political discourse, different approaches to thinking about wartime sacrifice are apparent: One (Canada) is focused on presenting a singularly seductive memory of a celebrated past to conflate militarism and nationalism in the making up of an uncontested Canadian identity. The other (that of the United States) is so preoccupied with enabling the war fighters' ability to wage full-spectrum dominance through a business model of efficiency and ownership of the future/potential realm, that it cannot carve out the time or space for the historical focus of body/memory politics. These differences in the representation of sacrifice, delineating a particularized scope and scale of sacrifice, can be made analytically useful only if analysed in tandem. These trajectories of sacrifice produce an effective tension in terms of how bodies might be differentially situated as agents of military violence and how that violence is socially organized, either through a departure from the citizen-soldier model of warfare (as in the United States) or through a symbolic celebratory reclaiming of the sacred grounds of violence (as in Canada). While Obama did cite the importance of sacrifice in his introductory statements, recall that it was presented in the form of a cursory, passing remark: Surely, sacrifice was something that happened in the past, but now that war is over, as shrinking defence budgets warrant, he does not imagine that embodied sacrificial actions will figure into future combat planning and operations.

In this vein of organizing violence through the rewriting of founding moments of the nation, a clear effect is the reasserting of the importance of sacrifice for the nation, which is seen in official wartime speeches, commemorative campaigns, and various annual days of remembrance. The oft-repeated question on days of remembrance, "How will you remember?" has a productive purpose. In being asked to re-remember certain historical wartime events, subjects are drawn into narratives and myths of national belonging and citizenship through interpellation, or hailing, to use Althusser's language. In the act of accepting these embodied sacrifices as just, subjects are reified as citizens while they are folded into a violent militarized identity.

In contrast, the US discourse reveals a desire for bodyless and bloodless war. Championed by the architects of the 2003 invasion of Iraq, particularly Donald Rumsfeld, the revolution in military affairs (RMA) seeks to replace military labour – that is, soldiers and, of course, the now politically problematic American casualties – with technology, capital, or "dead labour" (Parenti, 2007, p. 88). This desire is a function of contemporary expectations held by dominant political and military planners that war can be a phenomenon that, when sufficiently commodified, technologized, and exposed to neoliberal market logic and private-sector accounting and management techniques (Duffield, 2001, pp. 48–60), rendered efficient and "clean." This deliberate aesthetic construction of sanitized war not only contributes to the continued de-politicization of violence but also, more critically, offers the potential to make modern militarized sacrifice increasingly fraught. When war is ideally imagined as bodyless or bloodless on "our side," the space and meaning of military sacrifice is lost.

As Girard (1979) writes, regarding blood spilt in conflict, "the only way to avoid contagion is to flee the scene of violence" (p. 28). That people would actively avoid the contagion of the other by avoiding the stigma of bloodshed and death is further enabled and sustained through privatization and unmanned military technology, the logic extending one step further as soldiers do not even have to physically arrive onto the scene of violence but can apply, view, and analyse the violence and wreckage in a risk-free setting at a comfortable physical (and ethico-legal-political) distance. In the absolute prevention of contagion, there is no need to face the humanity of the other. The United States's preference for technological solutions (although framed in terms of human–machine teaming) has therefore shifted the terrain of sacrifice in that killing and being killed are no longer the legacy of the nation's citizen-soldiers.

As a collection of technological artefacts, strategic and tactical systems, and practices that supplement war fighters, the US DoD claims that ideally, these systems will gain greater autonomy such that "the algorithms must act as a human brain does." The DoD's (2014) report *Unmanned Systems Integrated Roadmap FY2013–2038* claims that "research and development in automation are advancing from a state of automatic systems requiring human control toward a state of autonomous systems able to make decisions and react without human interaction" (p. 68). While unmanned systems currently involve significant human interaction, the goals of net-centric warfare assume a gradual shift from humans "in the loop" to humans "on the loop," because of how it regards bodies as obstacles (Masters, 2008).

As a result, we are left with a paradox inherent to the current imagining of military sacrifice, expressed in the difficulty of how a government can position itself to simultaneously demand and disavow sacrifice. Turning on these political underpinnings of the demand and disavowal of sacrifice, this problematic also simultaneously teases out multiple tensions pertaining to fundamental political categories: citizens/others, national/post-national, local/global, peace/war, and public/private. For now, it can be said that militarized technological advancements, and the gradual supplementation of humans with technology such that humans ultimately sit further back "on the loop," would ostensibly allow for the potential of sacrifice, in its former nationalized and totalizing iteration, to be made obsolete. The effect of this is a sociopolitical shift in how violence is conceived and projected in accordance with ideals of the nation, expressed by the slow ejection of sacrifice from its historical legacy in the archetype of the citizen-soldier. The private contracting of soldiery, drones, and combat unmanning together capture a shift relative to the sacrificial logic that is bound up with how the nation harnesses the archetype of the citizen-soldier.

Still, despite the disavowal of sacrifice, the other side of the sacrificial coin (the demand for sacrifice) remains theoretically and analytically useful in analysing sacrifice in relation to a dialectical methodological approach. Sacrificial discourses, while perhaps appearing negated or sociologically irrelevant in comparison to previous political or historical moments, such as in the lead-up to, during, and after the Total or Great War, are also curiously and inconsistently *reasserted* (as in the Canadian examples) despite what I argue is the untethering of sacrifice from within the parameters of citizen-soldiering. Sacrifice is indeed currently demanded and appropriated by state officials, but I argue that its invocation – making up the geopolitical scene of sacrificial interpolation – is only effective when it draws on a notion of a historical past, and a specific narrative of sovereignty and sovereign authority therein, that is in fact *incongruent* with the architecture and practices of war today. In the avoidance of the violent scene of sacrifice through drone warfare, sacrifice itself then appears as a historical glitch, throwback, or anachronism. If the sacred is not "an idea but a presence" (Kahn, 2011, p. 147) then the appearance of the sacred – which must emerge in tandem with some axis of sovereign power wherein embodied combatants encounter one another on a battlefield – can be concealed just as it can be exposed. Either way therein exists a symbolic and political struggle, aimed primarily at governing a key archetype, the body of the citizen-soldier, to make the sacred (in)-visible.

Summary

A political-sociological account of sacrifice centres the body in sacrificial politics, asking what the social and political meanings of bodies might be when we account for their positioning and place in war. In contrast to dominant philosophical accounts of sacrifice, which tend to mystify, depoliticize, and/or romanticize sacrifice, I contend that how sacrifice is differentially mobilized or concealed across time and place tells us something about the materiality and changing social value of life and death in war. It is the mobilization or concealment of sacrifice – and indeed their coexistence – which gives sacrifice its puzzling quality. The paradox of military sacrifice, I argue, is exemplified by the struggle to both demand and disavow sacrifice and to manage the related contradictory trajectories of sovereign power and biopower therein. To maintain sovereign power, the traditional political–military capacity to wage war with appropriate weapons and soldiers much remain intact: Sacrifice must be demanded, and sovereign power must *take and consume*. However, at the same time, killing in war today must coexist with a biopolitics of protecting and saving (some) life through a disavowal or denial of sacrifice, which runs counter to sovereign power's need to consume. The simultaneous requiring and relinquishing of sacrifice coexists in a kind of dialectical synthesis, where these tensions and contradictions are imperfectly managed in ways that do not disrupt sacrifice *as a whole*. How this synthesis is governed, normalized, and resisted represents the crux of the investigation to follow. Historically, sacrifice justified and expanded violence but also operated as a check to limit it. As non-defensive wars are becoming progressively contested and ethically and legally indefensible, the associated forms of violence, while increasing in scope and frequency, risk being rendered meaningless absent a coherent understanding of sacrifice. Next, I explore how this tension between sovereign and biopower is seemingly, if not temporarily, resolved: in the governing of military sacrifice.

Chapter 3

Governing Sacrifice

The battlefield is strewn with the dis-emboweled and beheaded, with severed limbs and broken bodies. All have died a terrible death in a display of sovereign power. To view the battlefield is to witness the awesome power of the sovereign to occupy and destroy the finite body. It is to stand before the modern, democratic equivalent of the spectacle of the scaffold.

– *Khan (2008, p. 43)*

Making Sense of Sacrifice

I have argued so far that military sacrifice is not given in war. Yet we tend to think that it is, for historically justifiable reasons. Sacrifice is contingent – a product of political and historical contestation about the social meaning of sovereignty, and of bodies and death in war. It is both explicitly and implicitly deployed through historically specific framings of the nation as it relates to war's presumed place in society. According to Etienne Balibar (2011) states cannot become nation-states if they fail to consolidate this relation between the nation and sacrifice. Balibar writes that nation-states must *appropriate* the sacred, not only at the level of representations of a "sovereignty" but also at the day-to-day level of legitimation. Sacrifice in war allows for the appropriation of the sacred across both levels: at the level of sovereignty in addition to banal, day-to-day legitimation. I have shown how states drive this consolidation through political discourse. *Appropriation* implies contestation, and therefore politics. But this appropriation of the sacred in wartime, what I am suggesting here is about *making sense* of sacrifice, is increasingly fraught due to the changing quality and character of war.

On one hand, the temporally and spatially expansive quality of contemporary globalized war could in theory require conventional troop deployments and thus physical risk to soldiers; after all, technology can only do so much as a stand-in for human capabilities. This is even more apparent in that the United States has two major allies that are, at the time of writing, at war: Ukraine and Israel. However, the idea that soldiers might be deployed en masse in wars of choice and encounter physical risk is politically unpopular and undesirable in the United States and elsewhere. This is a historically specific tension related to a casualty-averse society at perpetual war: It must tolerate sacrifice in theory even as it discourages it in practice. The tension is illustrated when we imagine how sovereign and biopolitical power could possibly coexist in the context of war. These two forms of power have contradictory aims and objectives: One is about death, and the other is about life. This problem must be governed so that military sacrifice, at the level of the state and the everyday legitimation, makes sense vis-à-vis the nation-state.

How do states manage the competing demands of sovereign and biopolitical power? They render sacrifice *governable.* This chapter argues that the governance of military sacrifice in an era of biopolitics exploits a confusion between monarchical sovereignty and democratic sovereignty. For military sacrifice to make sense and survive alongside biopolitics, it must, on occasion, draw on certain aspects of an outdated *monarchical* sovereignty. This is evidenced, for example, by the Tomb of the Unknown Soldier, and the political and historical discourse and memorializing practices surrounding it, which I discuss later in the chapter. In highlighting some of the conceptual differences between premodern and democratic sovereignty, I claim that sacrifice risks conceptual incoherence within the political landscape of the democratic sovereign. As a result, this chapter concludes that, to gain credibility, sacrificial discourse today invokes aspects of premodern sovereignty to maintain the nation-state's grip on the sacred via notions of *legitimate* violence. This is a fragile process, which, as Bill Owens and Cindy Sheehan among others have pointed out, can quickly go pear-shaped, because legitimacy is not self-evident. Drone strikes are indicative of premodern sovereign power, a style of violence that tries to escape the constraints of embodied encounters in war while still maintaining a foothold on legitimate violence even as it betrays the humanness that gives it legitimacy.

This chapter proceeds as follows: I begin with a discussion of governmentality, claiming that for sacrifice to make sense, liberal governing manages it through market logic, expertise, and cultivating empowered subjectivities associated with the concept of freedom. After establishing sacrifice as a target of governmentality, I claim that part of this governing requires the reconciliation of sovereign and biopolitical power as they confront each other in contradictory ways. I discuss this in relation to the unfolding of neoliberal economic and political policy in the post–Cold War period. Next, I trace how this neoliberal restructuring of capitalism, which saw the expansion of US military bases worldwide combined with the downsizing of militaries globally, has enabled a new style of postmodern violence. This violence is characterized by privatization, technological fetishism, and the abandonment of the citizen-soldier model of warfare. The chapter then traces premodern and modern notions of sovereignty, asking how they might differentially inform, constrain, or enable sacrifice. The point in this section is not to uphold a neat distinction between premodern ("religious") and modern ("secular") periods[1] but instead to reveal how the neoliberal restructuring of capitalism, and the shifting notions of sovereign and biopolitical power therein, alter the content and meaning of sacrificial politics in global wars today. I draw on the case of the Tomb of the Unknown Soldier to illustrate this point, highlighting that to reconcile the tension between sovereign and biopolitical power, and thus to enable military sacrifice to make sense, wartime memory in the present is deployed to rationalize sacrifice, yet it contains fragments of anachronisms from the past that fit awkwardly in the present. Building on the question of what practices make speaking about a common object called security possible (Salter & Mutlu, 2012, p. 5), I ask what allows us to speak about sacrifice under liberal governmentality.[2]

Governmentality

Governmentality offers a framework to understand the application and operation of power beyond sovereignty. Foucault (1991) saw it as the 'art of government' and the 'conduct of conduct,' referring to control techniques and practices that are not reducible to state politics. Governmentality casts a wide net to understand how subjects are created, managed, and positively empowered to action by both state and non-state entities and actors across public and private domains. Governmentality thus gets us beyond sovereign power, or the law's capacity for spectacle, found in corporeal punishment and retaliation. For Foucault, the art of government emerged during the sixteenth century and consists of a twofold government of the self and government of souls, the government of children, the government of the state by the prince: "how to govern oneself, how to be governed, how to govern others, by whom the people will accept being governed, how to become the best possible governor" (p. 87). The essential issue for establishing the art of government is introducing economy into political practice: "To govern a state will therefore mean to apply economy, to set up an economy at the level of the entire state, which means exercising towards its inhabitants, and the wealth and behaviour of each and all, a form of surveillance and control as attentive as that of the head of the family over his household and his goods" (p. 92). In the sixteenth century, the economy signified a form of government, while in the eighteenth century, the economy became a level of reality and a field of intervention. The art of government means "the question of landed property for the family, and the question of the acquisition of sovereignty over a territory for a prince, are only relatively secondary matters. What counts essentially is this complex of men and things; property and territory are merely one of its variables" (p. 94). "Within the perspective of government, law is not what is important. This is a frequent theme throughout the seventeenth century, and it is made explicit in the eighteenth-century texts … it is not through law that the aims of government are to be reached" (p. 96). The good governor does not have to have a sting (weapon or sword); he must have patience rather than wrath – the positive content, which forms the essence of the governor and replaces the negative force. Power is about wisdom and not knowledge of divine laws, of justice and equality but, rather, of knowledge of things (p. 96).

The raison d'état acted as an obstacle to the development of the art of government. Yet the art of government had to contend with demographic expansion, the novel abundance of money, and the expansion of agricultural production, all through the emergence of the problem of the population. Therefore, the population becomes irreducible to the dimension of the family, and the family now becomes considered internal to the population as an instrument of government, rather than a model. As Foucault (1991) writes, "population comes to appear above all else as the ultimate end of government. In contrast to sovereignty, government has as its purpose not the act of government itself, but the welfare of the population, the improvement of its condition, the increase of its wealth, longevity, health, etc." (p. 100):

> In other words, the transition which takes place in the eighteenth century from an art of government to a political science, from a regime dominated by structures of sovereignty to one ruled by techniques of government, turns on the theme of population and hence also on the birth of political economy. This is not to say that sovereignty ceases to play a role from the moment with the art of government begins to become a political science … as for discipline, this is not eliminated either … in reality one has a triangle, sovereignty-discipline-government, which has as its primary

target the population and as its essential mechanism the apparatuses of security. (pp. 101–102)

Political economy is thus a central form of knowledge in governmentality. Governmentality, or control techniques and practices that are not reducible to state politics, does not indicate the loss of or departure from sovereignty but, rather, its reconfiguration within a triangle of power aimed at managing the population. Governmentality within our current neoliberal order must contend with a political economy that is both capitalist and globalized.

If the world is now understood through the economy, as primarily a marketplace, where market logic, capital, and expert knowledge now dominate political, social, and cultural life in a manner unprecedented in the history of capitalism, then it is no surprise that war, security, and its discursive and material counterpart – for example, sacrifice – are not exempt from the expanding logic of "neoliberal"[3] governing. In fact, war and security are coterminous practices explicitly harnessed by neoliberal ideology, for example, by reducing security to a commodity to be bought and sold in the global marketplace or encouraging former soldiers to take up employment in private military corporations. Key to an understanding of governing within neoliberal capitalism is how governmentality has *distance* baked in. Given its multidirectional and fragmented quality, governing is naturally decentralized, with no clear centre of power. In the context of global violence, this distance is emboldened through a combination of social, political, and technological means. Building on Foucault, in the following, I point to three interrelated aspects of governmentality that matter for our purposes.

The first relates to capitalism. In our current neoliberal era, capitalism, and its attendant market logic or rationality, is a defining feature of governmentality. In the context of war planning, this market logic does at least two things: It facilitates the investment, research, and capital necessary to create technological advancements in military capabilities to enable the vision of disembodied warfare, signalling that the protection of ("our") soldiers' lives the primary concern in war. Market logic under neoliberal capitalism also cultivates hyper-individualist, self-maximizing subjects. Soldiers are not exempt. This means that the ideal soldier subjectivity today joins an entrepreneurial-careerist mindset with notions of military service.

The second important point about governmentality relates to the notion of "patience rather than wrath." The ability to generate patience over wrath depends on the production of experts and expert knowledge required to enable and sustain this patience. To synthesize and implement the ideals of disembodied war, rational and deliberate calculation over life and death is valued over random, impulsive, or risky violence. The third point: Governmentality contains a tension between security and freedom (Ciccarelli, 2008) in that, to be secure, freedom must be managed.[4] The governing that results from this tension produces new socio-economic and political conditions for getting beyond the political challenges of military sacrifice. Because governing enables new subjectivities with *positive* associations, that is, the self-sacrificing citizen or citizen-soldier, it does not attempt to control but, instead, recognizes the capacity for action (control being an effect). It allows the discourse of military sacrifice to carry positive associations with freedom, enabling the subject to act as though free. The neoliberal, sacrificing citizen, for instance, spends their hard-earned money shopping per President Bush's infamous stick-it-to-the-terrorists recommendation. Joining the military and singing the national anthem at a sporting event (which may happen to feature military personnel) with renewed effort in the post-9/11 environment, for example, are meant to disable and resist terrorism and its impacts on freedom and in doing so have the effect of normalizing military interventions into countries whose citizens are deemed to be not free. Governmentality studies reveal how war is not purely destructive but also productive as far as it gives content and meaning to the nation-state while enabling a certain mode of governing and thus related kinds of action. War is productive in that it reconstitutes and normalizes the archetype of the citizen-soldier while calling on ordinary citizens to subscribe to militarization and identify as militarized subjects within and through discourses of history and sacrifice.

Liberal government contains two points of tension. First, "to maintain its own legitimacy, the state had to fashion the economy, civil society and the family as autonomous and self-generating and, at the same time, continuously and selectively intervene in these spaces to ensure their 'freedom' to function properly" (Brodie, 2008, p. 29). Liberalism presupposes that citizens regard themselves as individual bearers of sovereign power. Yet they nonetheless exist within a model of institutional pluralism where the task of liberal government is to construct and enforce pluralism while also developing "reflexive strategies of intervention to sustain this model," that is, social governance (Brodie, 2008, p. 29). The paradox of democratic legitimacy (Benhabib, 2004) implies that freedom must be managed. To this end, democratic legitimacy expresses a tension between "universal human rights claims and particularistic cultural and national identities" (Benhabib, 2004, p. 44). In the context of war or security, liberal government seems to require acts of sacrifice meant to preserve wider security.

Neoliberal governmentality embraces and amplifies the contradiction between security and freedom and makes subjection a condition of that freedom. The governing of sacrifice occurs at two levels, micro (subjectivity) and macro (nationally/institutionally). To be a subject willing to sacrifice the self and others to preserve freedom – a characteristically democratic ideal – is to begin to ease the burden between security and freedom. In this reconciliation of power, ordinary people are made complicit in normalizing war. The demand for sacrifice on the part of the state's military apparatus is where biopolitics (the administration of life) and the sovereign's support for militarization (the socio-economic and political planning for violence) correspond under the practices integral to neoliberal governing. Through the archetype of the honourable sacrificing subject, states more efficiently authorize legitimate violence abroad. Neoliberal governing reveals how war and security constitute new markets and new ways of being made free while paradoxically being *subjected to and subjectified by security.* Since the end of the Cold War, states have relied heavily on technology, and related forms of outsourced institutional, technological, and securitized expertise, to accomplish military and security goals related to full-spectrum dominance. In the next section, I apply neoliberal governmentality as a lens to explore the paradox of sacrifice in the post–Cold War era.

Governing in the Post–Cold War Era

The end of the Cold War ushered in a new economic reality, marked by uncertainty and economic insecurity owing to economic liberalization policies. These policies were adopted and promoted by dominant states, the International Monetary Fund, and the World Bank, which often tied loans to a receiving government's continued acceptance of austerity measures. This resulted in cuts to social spending, and thus a deepening divide between the global rich and poor, a "planet of slums" (Davis, 2006) combined with an epistemic crisis in the state system and neoliberal project (McMichael, 2009) writ large. The neoliberal restructuring of capitalism has a short yet aggressive history, initially coming to fruition via conservative coalitions in the United States and Britain in the 1980s. However, neoliberal reforms began earlier, in the 1970s. Following these reforms, private industries benefitted from several interrelated events, which coincided with significant reductions in social spending: The war on drugs, welfare backlash and severe attacks on welfare recipients, a shift within the prison system from rehabilitation to punishment, and the enhanced surveillance and criminalization of poverty, racial minorities, and immigration. Beyond North America, the collapse of the Soviet bloc and the subsequent downsizing of militaries worldwide, the privatization of state-owned industries across Europe, and the International Monetary Fund's and World Bank's endorsement and enforcement of these principles made privatization as a panacea appear inevitable (Avant, 2007).

According to Rose (1999), neoliberalism is not a simple return to nineteenth-century laissez-faire ideology. It was not a matter of "freeing an existing set of market relations from their social shackles, but of organizing all features of one's national policy to enable a market to exist, and to provide what it needs to function" (p. 141), namely techniques aimed at the responsibilization and entrepeneurialization of individuals, families, firms, and corporations. To govern better, one must govern less, ideally through the governing of the entrepreneurship of autonomous actors (p. 139). Thus, what emerged under neoliberal capitalism was the reconstitution of the relationship between the economy and the social: All aspects of social behaviour are now re-conceptualized along economic lines (p. 141) rendering the social and economic spheres antagonistic; fragmenting the social into a multitude of markets (p. 141). The division between employment and unemployment is now blurred when compared to the nineteenth century, which saw clearer divisions – moral, spatial, and economic. Perpetual insecurity is the norm for labour, from a global standpoint. The purpose of life itself under neoliberal capitalism is to produce a ceaseless economic capitalization of the self (p. 161). Active citizenship, as a result, is not defined by military participation or community service, for example, but by the extent to which one becomes an entrepreneur of oneself (p. 161).

In line with neoliberalism's privileging of the narrative of an unfettered market, security in this context is transformed from a public good into a commodity, packaged as a private service, delivered by private enterprise, and consumed by individuals as well as state and non-state actors (Avant, 2005). This included the privatization of public services and the abolition of capital controls, representing the foundation of "total capitalism" (Leys, 2008). Within total capitalism, the universal privileging of the standpoint of free-flowing transnational capital in the form of currencies or human labour, is matched by the unquestionable belief in the superiority of the universally applicable and uninhibited market (unburdened by historical contradictions or the complexity and diversity of culture). The rise of transnational corporations – many of which are much more influential than some states – shifted power away from democratically elected governments into profit-driven, mobile regimes aimed at managing the world as a globally assimilated economic unit (Clarke, 1996).

Total capitalism therefore favours the freeing of the market, market efficiency, hyper-individualistic competitiveness, and freedom of choice. Wendy Brown (1995) claims that privatization, a chief technique of governing at a distance, "violates public space, depoliticizes socially constructed problems and injustices, exonerates public representatives from public responsibility, and undermines a notion of political life as concerned with the common and obligating us in common" (p. 123). It was at this historical moment, Brown argues, that the discourse of "getting government off our backs" was amplified. Yet, simultaneously, there was also an aggressive consolidation of state power, as expressed by the state's very participation in deregulation, privatization, and contracting out its chief activities, identified as characteristically postmodern techniques of power (p. 18). Postmodern capitalism is therefore "monopolized without being concentrated or centered: it is tentacular, roving, and penetrating, paradoxically advanced by diffusing and decentralizing itself" (p. 32). Freedom – conceptualized here as synonymous with free markets and free choice – goes hand in hand with a reconfiguration of political authority towards decentralized and private means of governing.

Private means of governing redraw the parameters between the public and the private spheres (Hibou, 2004). While the national era produced a relationship between the state and citizens within national boundaries, the neoliberal era provides a significant challenge to the notion of the state as a seemingly fixed container. Yet, the public–private divide ought to be understood not as an objective fact but as a "signifying tool" used to make truth claims and is thus better analysed as a "complexity of shadows" (Memmi, 2002). To this end, there is no *pure* public sphere, just as there is no pure public one. Instead, it is the naming and representation of certain actions as public that attributes tremendous political power to that action. Conversely, naming something as private makes it invisible and individual. This means that privatization as applied to security (as a product of neoliberal rationality) views security not as a public good but as a commodity to be bought and sold within an uninhibited market. What does privatization do to state violence? Quite simply, it renders it private and thus invisible. Yet the state is a willing participant in these arrangements. An analysis of the state therefore must account for how states willingly offload power to various non-state organizations.

The United Nations is one example. The "United Nations [are to] establish peacekeeping missions and maintain international peace and security. They also establish 'order,' 'normalcy,' 'democracy,' and 'economic restructuring' as the goals of those missions" and "identify who is to be saved and who is to be left to die" (Whitworth, 2004, p. 33). The identification of "conflict-prone/third-world countries" is indicative of a colonial-era notion that advanced European states have a moral obligation to civilize others and that pacification is wrought through economic liberalization. So, "once illiberalism has been established as the problem, however, a series of responses to 'conflict-prone third world countries' presented themselves, regardless of whether those responses were embraced or resisted by local peoples, and regardless of whether they made sense within particular local contexts" (Whitworth, 2004, p. 43). Governmentality brings in state and non-state organizations to compel freedom, but this is a distinct brand of freedom associated with neoliberal capitalism and racially charged ideas about who is owed what in the global economy.

In the post–Cold War period, technological expertise, speed, and the hyper-reality of so-called bloodless war together illuminate the amalgamation of public and private powers to realize the goals of liberal wars of choice, wars conceived of in humanitarian terms – through a framing of protection – yet are nonetheless grounded in, and perpetuate, structural and material violence against people who are rendered politically, ideologically, and economically marginalized or otherwise considered ungovernable within the global economy. Bloodless wars are never bloodless or without suffering and loss. But the changing *meaning* ascribed to this bloodshed, suffering and loss (historically for the nation-state) gestures to an unstable, economy of sacrifice generated by competing forces, including nation-states, the military, private military forces, lobbyists, governments, and citizens. In the post–Cold War commitment to continuous technological advancements (to nurture the possibility of bloodless wars) the specificity of the violence therein is revealed: Violence, in its sheer fragmentation across public and private entities and its targeted, calculated applications, takes cover in hidden and discrete places, no longer seeking, or indeed requiring, recognition from a national audience. As will be argued later, rather than a sense of ownership in public demonstrations of violence, drone violence denies itself in the moment of its enactment (Rupka & Baggiarini, 2018). This style of violence is specific to postmodern war: Military capability is expanding within neoliberal capitalism, while social and political norms that have traditionally centered humans and a form of political authority associated with the nation-state in war are collapsing.

The Violence of Postmodern War

The violence of postmodern war exists against the backdrop of the neoliberal restructuring of capitalism in the post–Cold War era. This period witnessed a process of increasing militarization of technology and the expansion of military bases worldwide supported by expert knowledge, market logic, and neoliberal discourse of freedom. At the same time, militaries worldwide were downsizing, thus recruiting, training, retaining, and deploying fewer soldiers. Put differently, military capability was widening and deepening in scope, as troop numbers decreased. While some might argue that this was because the looming threat of war between East and West had ended, as signalled by the triumph of neoliberal capitalism, Whitehead and Finnström (2013, pp. 1–25) invite us to think about not war's *end* but, rather, its changing character. To this end, they suggest that the global restructuring of capitalism's relation to war reflects the reconfiguration of a particular style of violence. This violence manifests as cyclical outbursts without guaranteed endpoints that defy traditional spatial and temporal limits of violence. A "war on terror" is ideological and as such can never really end; its logic knows no bounds. "We" ("civilized society") can never overcome terror, so likewise, no parallel strategy leads to a traditional victory or an end to violence. Whitehead and Finnström argue that

> war is not an aggregation of violent acts that finally reaches a given threshold of carnage to become war rather than armed conflict, counterinsurgency or peacekeeping, but is the invocation and creation of a particular political reality that engages allies and enemies, civilians and soldiers, in a particular style of violent interactions. (p. 7)

What is the nature of this political reality, and style of violent interaction, that is made possible through patience rather than wrath, technological prowess, and casualty aversion? Larry George (2002) claims that war in the post–Cold War era serves to medicate and pacify such social ills. President Obama's approach to war, which did not require a national audience, was driven by singular, repetitive, yet isolated and frequently invisible strikes and assassinations as a preventative form of healthcare for the political body. As George explains, the protracted global war on terror reveals war as paradoxically both the disease of and remedy for the social body. These "pharmacotic" mimetic wars use sacrificial rituals in a "politically cathartic and unifying function- to 'cleanse' and 'purge' these societies of internal disorder and remove troubling dissenters" (p. 164). This quelling of internal disorder, a parallel move alongside foreign wars and occupation, is evident in how military technologies are used domestically to exploit the boundary between law enforcement and war. George further claims that war sanctifies politics by transubstantiating the blood shed by compatriots, enemies, scapegoats and dissenters into various kinds of political power. The excess of sacrificial violence – what, in Giradian terms, would amount to the overwhelming intensification of mimetic violence, resulting in a violent contagion and the direction of violence "against all" to "against one" – serves as political fuel to authorize renewed violence.

But what if the centre of violence is not immediately known or even visible? Who or what entity is cleansing or purging, being purged or cleansed? George's account therefore takes the publicity of sovereign power for granted, overlooking its character in democratic societies. In contrast to the monarchical sovereign, which George's analysis implicitly invokes, the democratic sovereign, which characterizes our modern period, is immanent but largely invisible (Singer & Weir, 2006, p. 445). This sovereign right to execute through drone strikes is masked by a regime of disciplinary and panoptic power, which operates "through a regime of visibility leaving the center of power concealed" (Singer & Weir, 2006, p. 445). If the source of that sacred power is concealed, particularly when nations and their audiences are not explicitly called on or folded into the sacrificial system, imagining what the purpose of pharmacotic war is or what kinds of fungible political power are available to be captured and by whom becomes difficult. The neoliberal privatization and fragmentation of social relations have altered the content of sacrifice such that those who authorize sacrifice, and what entity precisely is meant to receive the surplus or benefits, are not immediately known.

Consider the language used by George: "sacrifice," "ritual," and "transubstantiation," words that conjure up the notion of the sacred even as his pharmacotic metaphor is strictly secular. This reveals the problem of failing to distinguish between premodern and democratic sovereignty, where, traditionally, religion (and God's monopoly on truth) is associated with monarchical sovereignty and secularism with democratic ideals and revolutions. If the sacred is most immediately aligned with an anachronistic form of sovereign power, then we must account for how the sacred survives in a secular political setting, where truth is pluralized and thus knowledge about the content and meaning of sacrifice is obscured. I suggest in the final section of this chapter that memorialization and memory practices around war offer a clue into how premodern sovereign conceptions of the sacred manage to continue to circulate politically within democratic sovereignty, such that sacrifice continues to be linked to the justification and authorization of nation-state violence, despite the theoretical and empirical evidence of a weakening connection between agents of state violence – citizen-soldiers – and sacrificial cults.

If war is not to engender national boundaries and demarcations of populations and social relations therein, and if it does not call upon a national audience, then what is to be made of the sacrificial excess? A nation that does not call on its citizens for war is not categorically a nation at all (Taussig-Rubbo, 2011). Postmodern violence (cyclical outbursts with no guaranteed endpoint) invokes Sylvester's (2012, p. 489) argument that war does not create new social relations per se but, instead, exacerbates social tensions that are already present. She writes that "wars are not indicative of a breakdown of inter-state balances of power, norms or structures of conflict resolution as much as they are an *intensification* [emphasis added] of social relations" (p. 489). The qualities of these social relations, characterized by forms of violence imminent in society[5] and intensified by abandoning the citizen-soldier model of warfare, and drone warfare – privatization, fragmentation, and the destabilization of the category of the witness – are explained later in the book. For now, I return to Foucault's conceptual map to further elucidate sacrifice on the continuum between sovereign and biopolitical power.

Foucault's work shifts the emphasis away from the intentionality of states to their techniques and apparatuses of regulation. He claims that in the eighteenth and nineteenth centuries, when the domain of the biological came under the explicit control of the state, biopolitical technologies of power were manifested within and against the population. The maintenance and administration of the population via the family represents the crux of the biopolitical problem. Biopolitics is meant to enhance the equilibrium in society through a disciplining of individual forces in conjunction with regulatory technologies aimed at the biological characteristics of the population: in other words, for the first time, the population was rendered a political problem, and the family was one entry point into the population. Populations' overall health, longevity, rates of reproduction, fertility, illness and mortality, hygiene, sanitation, and the concern over epidemics become targets of power and calculability under the rubric of the state's newfound knowledge: statistical analysis (Foucault, 1978).

This biopolitical mandate informed the making of war such that the preservation of life was its impetus. In an oft-quoted passage, Foucault remarks that "wars are no longer waged in the name of a sovereign who must be defended; they are waged on behalf of the existence of everyone; *entire populations are mobilized* [emphasis added] for the purpose of wholesale slaughter *in the name of life necessity* [emphasis added]: massacres have become vital" (Foucault, 1978, p. 137). For Foucault, once the former right of the individual sovereign has been replaced by the right of the social body to maximize its health, the question of war similarly shifts to become, "How can the power of death, the function of death, be exercised in a political system centered upon biopower?" (Foucault, 2003, p. 254). Foreign – inspired and legitimized by both imperialism and racism – reflect the beginnings of an answer.

Foucault (2007) argues that race became a biopolitical paradigm in the seventeenth century when war emerged as the permanent basis for all institutions of power. War became both a model and a principle of intelligibility of politics and thus a grid for analysing politics. National narratives of war and articulations of national consciousness entailed a related understanding of "foreign" enemies, a narrative that extended the legitimacy of the nation and naturalized violence against perceived others while affirming its territorial borders and need to protect the nation from "dangerous" internally produced and externally threatening peoples. The image of the foreigner is a productive artefact of national thinking about racial otherness and is central to the configurations of citizenship. As Achille Mbembé (2003, p. 16) claims, power makes continuous use of the state of exception discourse, notions of emergency, and a fictitious enemy. For Mbembé and Foucault, in the economy of biopower, the function of racism is to regulate the distribution of death and to make possible the murderous functions of the state (Mbembé, 2003, p. 17). Recalling Foucault's earlier question about how death can function in a system centred on life, it appears that racism offers a clue.

Racism perpetuates the notion that certain people are inherently inferior, unequal, or backwards; ungovernable; or unable to sufficiently self-govern within a liberal economy. This belief system is fundamental regarding the question of how death can quietly function in a political system centred on life (Foucault, 2003, p. 254). Historically, abhorrent racism can be seen in how the National Socialist Party systematically exterminated millions of Jewish and other marginalized peoples during the Holocaust (although extreme right-wing, neo-Nazi, and "immigration reform" based parties are quickly re-emerging in many parts of Europe and North America). Nations, as cultural conceptualizations,[6] are imagined as ethnically pure entities. National belonging then implies a dialectical relation with its opposite: exclusion and violence against perceived others. Nationalism at its extreme demands a violent relationship with the other where ethnicity is used to conceptually organize insiders from outsiders. Within affirmative performances of national belonging, there is also a symbiotic way that differences, exemplified by the social and political marginalization or outright exclusion of gendered and racialized others are simultaneously invented and put into sharp relief in relation to that very act of the performance of national belonging.

As Achillé Mbembé (2003, p. 23) explains, the "synthesis between massacre and bureaucracy, the incarnation of Western rationality" witnessed the simultaneous demarcation of the colonies as a space of exception where the rules about sovereignty and confined warfare among equal states did not apply. While Mbembé argues that the state undertook ways to civilize and rationalize the sovereign act of killing, it did so under the guise that "the centrality of the state in the calculus of war derives from the fact that the state is the model of political unity, a principle of rational organization, the embodiment of the idea of the universal, and moral sign" (p. 24). While Mbembé is sympathetic to Foucault's use of biopolitics, it is in this context that Mbembé questions whether Foucault's analysis is conducive to a practical analysis that uncovers these colonial and post-colonial relations of enmity (p. 12). As such, the racial denial of any common bond between the conqueror and the native (p. 12) was used to justify colonial war and occupation – a matter of "seizing, delimiting, and asserting control over a physical geographical area ... the writing of new spatial relations (territorialization) was, ultimately, tantamount to the production of boundaries and hierarchies, zones and enclaves; the subversion of existing property arrangements, the classification of people" (p. 26). The processes of the exteriorization, rationalization, and spatialization of war included the identification of colonies and their racialized subjects as zones of exception and as such the impetus to kill was realized outside, or as "exception" inside the rules contained within Western legal and moral achievements.

Sovereign power, which Foucault defined as embodied by the king, is a repressive, spectacular, and prohibitive form of power. In contrast, biopolitical power is a productive power that aims to enhance life by fixing on the management and administration of life via the health of the population. It compels people to act in particular ways – as consumers, patriots, or sexual citizens, for example – to self-govern. Hence, it is a tool that allows for "governing at a distance," because the origin of that power is not traceable back to a single source. Foucault's model of power does not give concrete answers in terms of isolating power's precise location and origin but instead offers *an approach* to thinking about the materiality of sacrificing bodies,[7] and the function and meaning of the power of war and militarization (the impetus towards death) within a society that claims to be founded on the democratic principles of life.

Furthermore, the tension wrought by militarization, put in Foucault's (2003, p. 241) terms, in liberal democratic society today, goes to the heart of the paradox of liberal modernity insofar as how it views life: Liberal government is concerned with how to satisfy the conditions for two competing types of power to coexist: the first being sovereign power, "to take life or let live," while the second is biopolitical power, which fosters life-enhancing techniques and programs, to "foster life or disallow it to the point of death." The former sovereign right was to "take life or let live" (p. 241); however, states became increasingly concerned with regulating and optimizing the health of populations, which co-emerged alongside the development of social scientific disciplines generating new knowledge about populations through statistics, psychiatry, and medicine. These disciplines proffered proper conduct, for example, for children, including how to respond to and control their sexuality, and related expert knowledge around key figures of moral regulation: the sexual deviant, the hysterical woman, the criminal, and the madman. These subjective figures all coalesced within a framework of biopolitical governing. Thus, biopower came to focus on the body in terms of the individual – what Foucault termed "anatamo-politics"[8] or the maximization of the individual body and the collective, the biopolitics of the population as a whole. The body served as a target of power organized around promoting and enhancing life, while at the same time, the destruction of the body is evidenced by sovereign power's investment in spectacles of violence (Foucault, 1978).

These contradictions of militarization – how a liberal society can claim the belief that life ought to be protected, but have entire industries and political-economic policies devoted to killing – further unfold in the management of the space between life and death, or the demarcation of politically important life, as opposed to non-political, or what Agamben (1998) calls "bare," life. Agamben criticizes Foucault by extending the biopolitical moment, when politics seized hold of life, further back in history to the ancient Greek polis. He claims the production of a biopolitical body is the original, albeit concealed, activity of sovereign power. He argues that modernity is predicated on processes of demarcation, regulation, and discipline, which render some expressions of life politically meaningful: life that is regarded as transcendent and proficient is marked by its political capacity. This political life contrasts with mere animal existence (bare life) that is marked by a lack.

According to Agamben (1998), biopolitics is intensified today, as evidenced by the camp – the most absolute biopolitical space to have ever been realized. Biopolitics entails classifying morally and politically qualified life (bios) and bare life (zoe), taken from Aristotelian classifications. These delineations are based on the public sphere as equated with the political and the private sphere with the biological. For Agamben, the concentration camp is the apotheosis of this pure and absolute biopolitical space, where bare life meets sovereign power. Critically, for Agamben, there is "no return from the camps to classical politics" (p. 188). All life has the potential of being rendered bare life, or the non-citizen subject of politics. Of the camp, Agamben (1998) states,

> If this is true, if the essence of the camp consists in the materialization of the state of exception and in the subsequent creation of a space in which bare life and the juridical rule enter into a threshold of indistinction, then we must admit that we find ourselves virtually in the presence of a camp every time such a structure is created, *independent of the kinds of crime that are committed there and whatever its denomination and specific topography* [emphasis added]. (p. 174)

Within the camp, distinguishing between fact and the rule of law is impossible, and this is reflected in the camp's property of immanence whereby the camp as a state of exception becomes the rule but, as an exception, remains nonetheless outside the normal order (Isin & Rygiel, 2007, p. 183). Isin and Rygiel (2007), to be sure, build on Agamben's analysis of the camp as a demarcated space invoking unique juridical procedures and deployments of power over bare, politically incapable, or unqualified life. Arguing that Agamben's conception of the camp is paramount to theorizing the relations between qualified and unqualified life, they claim that it is nevertheless both essentialist and ahistorical. They suggest, rather, images of "frontiers" and "zones" as spaces where subjects are processed as inexistent beings – non-citizens in waiting. They write:

> If the camp was a space of abjection where people were reduced to bare life, the zones, frontiers, and camps of our times are abject spaces, spaces in which the intention is to treat people neither as subjects (of discipline) or objects (of elimination) but as those without presence, without existence, as inexistent beings, not because they don't exist, but because their existence is rendered invisible. (p. 183)

Isin and Rygiel complicate Agamben's notion of the camp to include questions of geopolitics and space and by broadening the scope and scale of the logic of the camp to include the possibility that those exposed to its rationale might, in fact, be non-existent/invisible and therefore do not even qualify as subjects/objects to be made bare. Indeed, these kinds of contextual specificities that these critiques call for suits a Foucauldian rejection of theoretical closure in line with a political-sociological view of sacrifice. Yet the question remains as to what relation biopower entails to the sacred. In this vein, it can be said that since sacrifice resides in the domain of killing (sovereignty) biopower's relation to sacrifice can only be indirect. The biopoliticization of war ("wars are waged on behalf of *everyone, in the name of life necessity*") (Focault, 1978) has witnessed biopower and biopolitics taking many forms in contemporary war-making. Consider it in the form of so-called humanitarian wars or in the process of military privatization and/or the integration of drone technology into the battlefield: All see the protection of life as their primary concern. Biopower then acts as a kind of counterforce to sovereignty and on the application of sacrificial politics that shores up sacrificial politics' incongruity with postmodern violence. Agamben's analysis shows how the logic of the camp sought to reduce humans to *Homo sacer* so that they could then potentially be made bare: identified, processed, tortured, and exterminated.

Yet, drone strikes *skip these steps outright*. Cultivating the space to contain bodies – to identify, detain, torture, and process bodies – appears to be no longer the primary concern of fatigued American military planners as it perhaps once was in the immediate post-9/11 era. Consider the following interview excerpt from a lawyer working directly on issues pertaining to national security and targeted killing:

> The alleged legality [of drone warfare] is under the President's war powers. As Executive and commander in chief of the U.S. military, and specifically under the Authorization for the use of military force that was issued after 9/11. The Authorization for the use of military force is a four-line law; it essentially authorizes lethal action against anyone who is supporting or was supporting an arm of al Qaeda in connection with the 9/11 attacks, which is primarily where we attack them, because now, you know, however many years later the connection between who they're killing now and what happened on September 11th are pretty shaky." (SSB, Interview)

That killing might occur on "shaky" legal, political, or ethical grounds reveals how and why drones and drone strikes might be deployed to overcome these constraints. The political and ethical failures of the Guantanamo Bay and Abu Ghraib prisons remind us of the contradictions between biopolitical and sovereign power in the exercise of post-9/11 security: Indefinitely holding, processing, and torturing bodies are wrapped up with a responsibility for the well-being of those same bodies. The logic underpinning a drone strike is to dislocate the violence of the camp, bypass its spatial and geopolitical constraints, and enact disembodied force and thus the embodied ethical responsibility associated with it.

North Waziristan, Pakistan, is one exceptional site of governance among several where violence (understood here as a broad continuum of acts from permanent surveillance to killing) is externalized onto the bodies of racialized others (Coll, 2014; Tahir, 2017). Here, the rules of conventional warfare per international humanitarian law do not apply. The subjects of drone surveillance and strikes made up a community of perpetually terrorized victims in negotiations between the United States and its allies or enemies during the peak of the global war on terror. Drone technology is hailed as that which will save us from (or at least sanitize) our violent, collective sins. Drones are presented as the solution to the spatial, political, and economic limitations and long-term political unworkability of the physicality of camps – detention centres like Guantanamo – or the logic of Guantanamo applied to air-borne technology such as drones. Extermination through targeted killing enables this policy without the immediate political risk, evading the humanity of the other.

In this evasion, biopower has not displaced or surpassed sovereign power but, rather, is redistributed in a field of governmentality. Sovereignty is not a static concept but a set of practices aimed at including and excluding, creating boundaries and limits, and shaping political agency (Connolly, 2004). It is described as an "animated anachronism" (Butler, 2004) and in terms of a "migration."[9] It is thought to have produced petty sovereigns outside the nation-state,[10] revealing how globalization and corporate interests exacerbate a "crisis of sovereignty."[11] This means not that national sovereign power is disappearing but that it is being reconfigured[12] insofar as the state's relation to providing citizenship entitlements thus also in ways that leave the citizen-soldier's relation to sacrifice hollow. Importantly, the devitalization of sovereignty does not always coincide with the emergence of governmentality (Butler, 2004, p. 53). As I discuss next, governmentality was predicated on the shift between premodern sovereignty and democratic sovereignty and the shift of symbolic regimes therein (Singer & Weir, 2006/2008).

The Democratic Sovereign

In highlighting some of the conceptual differences between premodern and democratic sovereignty I claim that sacrifice risks conceptual incoherence within the political landscape of the democratic sovereign. As a result, this chapter concludes that, to gain credibility, sacrificial discourse today must invoke aspects of premodern sovereignty to maintain the nation-state's grip on legitimate violence. Consider the following description:

> The battlefield is strewn with the dis-emboweled and beheaded, with severed limbs and broken bodies. All have died a terrible death in a display of sovereign power. To view the battlefield is to witness the awesome power of the sovereign to occupy and destroy the finite body. It is to stand before the modern, democratic equivalent of the spectacle of the scaffold. Viewing the battlefield from a certain distance, it is not even clear who is the object of sacrifice: the enemy and the conscript suffer the same threat and burden of physical destruction for the sake of making present sovereign power. (Kahn, 2011, p. 43)

Kahn's seductively written yet historically ambiguous imagining of sovereign power's place on the battlefield offers mixed metaphors for sovereign violence. As his description slips between premodern and modern ideas of sovereign power, in the opening two sentences, we are very much reminded of the tremendous and spectacularly crushing randomness of monarchical sovereign violence, while the third sentence effortlessly cites the imagery characteristic of modern democratic sovereignty. It appears that from this powerful yet historically incoherent description of the relation between sovereignty and sacrifice we cannot give sacrifice the nuanced treatment required to export its theoretical value into our empirical context.

In this vein, Khan asserts a given or static quality of sovereign power. Its ostensible immediate publicity, as emerging in the violence that flows from one body to the next, the sovereign makes itself present in that it *occupies* these bodies through their conflicted association as both citizens (bound by duty) and individuals (bound by free will), making himself visible as they fight to the death. In Kahn's portrayal of sovereignty, there is an implied invitation to view and experience (to gain knowledge of) the public spectacle of violence. Khan's narrative therefore has the effect of placing the body of the conscript and the body of the enemy in a shared space as counterparts, as far as they are both equally and simultaneously exposed to and compelled to embrace the destructive power of the sovereign as evidenced in the violent encounter. This battlefield assumes the presence of bodies where sovereignty is realized in their embodied violence with each other, in their encounter of destruction and subsequent bloodshed. While Kahn is not explicitly discussing battlefields particular to the expressions of sovereign and the biopolitical power contained in drone warfare, it is nevertheless productive to export his description of the battlefield to supplement the question of sacrifice, sovereignty, and biopolitics. Yet, if we were to supplement Khan's analysis by granting sovereignty its necessary historical parameters, different symbolic regimes and therefore different relations between knowledge, power, and law, would be revealed in the distinction between pre-modern and democratic sovereignty.

Forming a "famous glittering triangle," Singer and Weir caution Foucauldian scholars of the danger in treating sovereignty as a contrasting concept in relation to discipline and governance (where the terms are posed antithetically) rather than as an analytically distinct concept and therefore productive in and of itself. Singer and Weir, whose work is worth summarizing at length, argue that historians of the present inherit this tendency from Foucault in that they do not adequately treat sovereignty as a concept worth investigating but, instead, use it as a straw man to bring into focus different notions of power.

My argument in this section is that the shift between monarchic to democratic sovereignty is critical to understanding how sacrifice is lodged in the archetype of the citizen-soldier, supported by the tradition of democratic republicanism. Following Singer and Weir (2008), historical parameters are necessary to better understand the dangers inherent to the conceptual fusion of sovereignty and governance. Moreover, a historical focus reveals how changes in the logic of sacrifice are tied to the underlying change in symbolic regimes. I claim later that despite its locus in democratic sovereignty, remnants of premodern sovereignty circulate at the site that is the Tomb of the Unknowns. The presence of anachronistic iterations exists to reaffirm the importance of sacrifice in preserving the transcendental qualities of nation-state violence. For the ritual of military sacrifice to remain credible, considering the contemporary style of warfare, which attempts to eschew embodied violence, it must occasionally draw on anachronistic notions of sovereign power. The Tomb of the Unknowns then represents intersections between premodern and democratic sovereignty – the "secular" and the "religious."

According to Singer and Weir (2006, pp. 451–452), Foucauldian scholars insufficiently theorize sovereignty in four ways: first, on the issues of historical specificity – they claim that sovereignty came into wide use during the Renaissance, not the medieval period; second, on the false equations of sovereign power and law – sovereign power and law ought not be equated as they have different temporalities. Laws are meant to last, while sovereign power always bears an excess relative to the law, as exemplified by the state of exception (the "temporary" suspension of constitutional law under "emergency" conditions); third, sovereign power's false equivalence with the state; and, fourth, sovereign power's false equation with a delimited territory: It was only when sovereignty became "horizontal" that it was properly territorialized. Singer and Weir argue that, by giving governance an exterior, by extension, so, too, would sovereignty reclaim its conceptual and historical parameters.

However, because of this conceptual incoherence, sovereignty, in Foucauldian literature, becomes a residual category because it is used only superficially to draw a comparison to discipline, or governance. Additionally, this dearth of conceptual rigour encroaches into governmentality theorizing, whereby governance is often conflated with politics and the state; governance then ends up conflated with all forms of power. Although the early Foucault was largely disinterested in sovereignty (the archaeological writings did not foreground sovereignty thematically), he still nevertheless, according to Singer and Weir, established some defining features of sovereignty.

Foucault argues that sovereignty concerns itself with the juridical subject, discontinuous extraction of taxes and labour, and the enactment of law. Sovereignty revels in spectacular displays of the monarchical sword. Its object is the land; its power is exercised over a territory. Furthermore, Foucault claims that sovereignty was a central form of power prior to the modern era, in which it is associated with the state, articulated in terms of law. Its preeminent form of expression is the execution of wrongdoers. However, much of this theorization of sovereignty is communicated not through a direct study of the concept itself but in relation to its antithetical form of power – discipline. Discipline, conversely, is aimed not at juridical subject. It enacts continuous surveillance of individuals, acts through standardizations set through technically established norms, and operates through a regime of visibility leaving the centre of power concealed, and its object is the body (Singer & Weir, 2006, p. 445).

Foucault maintained that the premier instrument of sovereign power has been law. The monarch symbolized the living body of sovereignty, while power possessed a single centre. The law was the expression of that centre's will. Where the theory of right tried to limit monarchy to match it with legitimacy, juridical systems later democratized sovereignty by linking it with public right, resulting in popular sovereignty. Yet Foucault argued whether monarchical or popular, sovereignty was still reducible to law, and as such, he advocated eschewing an examination of the genesis of sovereignty in favour of an examination of domination via the manufacturing of subjects. Hence, the impulse in Foucault, and Foucauldian scholars, is to study power apparatuses and power effects rather than documenting who has power and what their intentions might be.

For Singer and Weir, perhaps the most egregious failure of Foucault and his descendants – including Butler, whose view that sovereignty has resurged in the field of governmentality problematically builds on and extends Foucault's absence of a genealogy of sovereignty, and Agamben (2008), whose historically deterministic ontology of sovereignty treats bare life as "part of the inevitable tragedy of Western political thought from its inception, rendering the state of exception almost inevitable" (p. 65) – is the ahistorical presentation of sovereignty as seen in the lack of distinction between monarchical and democratic sovereignty. This in-distinction, or conflation, is further exacerbated by the abandonment of the symbolic after the period of the French Revolution. It is in this disregard of the symbolic that limits the understanding of democratic sovereign, as well as how the "symbolic regime," to use their interpretation of Claude Lefort's language, shifted to accommodate governance.

What Foucault terms governance depends on this shift in the symbolic regime – the division between power, knowledge, and law. Drawing on Lefort, Singer and Weir (2006) argue that changes in sovereignty rely on changes in the symbolic regime:

> For the claim here is that the presentation of an orderly, meaningful world in common that extends beyond restricted localities is very much dependent on how power, knowledge, law and their interactions are themselves presented and represented. The very sense of the most basic distinctions through which a common world is expressed – those separating the real from the unreal, the natural from the artificial, the orderly from the disorderly, the just from the unjust, etc. – is conditioned by the articulation of knowledge, law and power within a given symbolic regime. (pp. 452–453)

Governance supposes a certain secularization that renders both power and knowledge immanent to this world. It supposes a separation of knowledge, power, and law, and its attempt to articulate the two terms is conditioned by that separation (p. 457). To better comprehend the meaning of the symbolic and sovereignty, the authors contrast the symbolic regime within the ancien régime and that characteristic of the democratic revolution.

In the ancien régime, knowledge, power, and law are acquired, and converged, through God. The figure of the king made the divine will present by representing it, as he is the ultimate embodiment of all law. "He presents the 'scene' in which power, as the source of order and justice, can be seen and be seen to be present" (Singer & Weir 2006, p. 453); thus, "representation does not represent what is already present but gives presence to what is not present because otherwise invisible" (Singer & Weir, 2006 p. 453), and as such, in the case of the monarch, "symbolic power is the direct effect of the claim to be the visible intermediary of an invisible, transcendental realm. Thus, the importance of his presence, words and acts: his is not a mere presence as it exists on a higher plane of reality; his words are not mere words, as they mold the world of things; and his acts are most often ritual acts because they serve to communicate between worlds" (Singer & Weir, 2006, p. 453).

In contrast to the monarchical sovereign, the democratic sovereign is immanent but largely invisible. This is because "the people" is a fleeting, unstable, and uncertain category. The democratic sovereign includes a symbolic regime wherein knowledge, power, and law are divided. After the democratic revolution, culminating in the eighteenth century, sovereignty no longer relayed a transcendent order but was imminent to the newly emerging concept of "society," thereby severing power's link to the supernatural and natural worlds (Singer & Weir, 2006, p. 454). Singer and Weir (2006) write that the heterogeneity of power – truth regimes are now pluralized – witnessed proliferating knowledge-holders: intellectuals and experts – shattered the Chain of Being as far as the laws of nature were no longer regarded as having any bearing on juridical laws. Thus, "the sovereign people may now be immanent to society, but this sovereign does not exist positively, except in privileged moments – revolutions and elections" (p. 455). The sovereign is less a positive reality than a barely visible referent, because it is never entirely clear what the people want or even who they are. Unlike the monarchical sovereign, who provided a sense of certainty, the democratic sovereign cannot provide certainty because the visible powerholders, who are on the same register of reality as individual citizens, are always subjected to antagonisms in the form of questioning, contestation, and legal overthrow, weakening their relation to truth. This leaves the seat, or place of power, as the authors claim through drawing again on Claude Lefort, *empty*.

What happens to military sacrifice when the seat of power is empty? A discussion of the Tomb of the Unknowns, or as it is more colloquially known, the Tomb of the Unknown Soldier, clarifies how the shift in sacrifice's tie to the citizen-soldier is underpinned by the shift from a monarchical to a democratic sovereign and thus the related shift in symbolic regime described above. However, the Tomb of the Unknown Soldier is a space where premodern sovereignty is preserved and memorialized – therefore, I argue that notions of premodern and modern sovereignty are not opposed but instead intersect in the creation and preservation of such a space. This intersection reveals how sacrifice maintains its hold in nationalized violence. To justify the possibility of military violence, states must maintain a historical past (however fictitious) wherein the ordinary and profane realms can be momentarily synthesized, transcendental ideals can be positively embodied, known, and communicated. Death for the survival of the new, secular religion – the nation[13] – *must be* meaningful. If sacrifice were to lose this connection to the symbolism of premodern sovereignty it would fail to connect the nation-state's violence to the future health of society. Its citizens, the people, as an abstract concept, would not be immanent but would instead risk fading away into an antipolitical landscape characteristic of the ancien régime.

The Tomb of the Unknown Soldier

Recall Singer and Weir's (2006) claim that, in democratic sovereignty, "the sovereign people may now be immanent to society, but this sovereign does not exist positively, except in privileged moments – revolutions and elections" (p. 455). Military sacrifice also enables the positive experience of the sovereign. The Tomb of the Unknown Soldier and attendant calculated memorialization[14] practices also confirm the positive existence of a historically anachronistic sovereign power. Therein, we see the confirmation of the constructive interrelation between the two historical contingent manifestations of sovereignty in the production of shared meaning, and of how social memory utilizes war as a site of collective identity.

Located at Arlington National Cemetery in Washington, D.C., the Tomb of the Unknown Soldier is a popular monument dedicated to American service members who died unidentified. The first Unknown was a First World War soldier, while crypts of the Unknowns from the Second World War, Korea (added in 1958), and Vietnam (added in 1984) are featured west of the First World War Unknown. The fact that the soldiers are unidentifiable confirms Lefort's claim that the seat of power in democratic sovereignty be empty: These tombs may as well be empty, because their contents are vacant of identity. Thus, "the people" fill the tomb as ersatz soldiers, symbolically sharing in the sacrifice of those who died in war. Sculpted into the east panel, which faces Washington, D.C., are three Greek figures representing peace, victory, and valour. Inscribed on the back of the tomb are the words *"Here rests in honored glory an American soldier known but to God."*

The "known but to God" phrasing is typical for a premodern articulation of sovereignty. Recall that in the *ancien régime*, knowledge, power, and law consolidated in the body of the sovereign authority who made visible, made present by representing, what would otherwise be invisible: divine, transcendental truth. Thus, viewers at the Tomb are told that there is some element of these sacrificial deaths that can only be interpreted by one (heavenly) body. This depoliticizes sacrifice in that by mystifying sacrifice it works to encourage people to submit, without much questioning, to a higher order, to cosmological truths as can be told only by God or his human representative. Referring briefly back to Singer and Weir (2006): "He [the sovereign] presents the 'scene' in which power, as the source of order and justice, can be seen and be seen to be present" (p. 453); thus, "representation does not represent what is already present but gives presence to what is not present because otherwise invisible" (p. 453). For these sacrifices to be made visible, they must be represented. The notion that only God can do this representing dramatically depoliticizes wartime deaths and suffering, making sacrifice a normal part of nation-state violence.

Consider also Ronald Reagan's speech at the Tomb of the Unknowns, on 11 November 1988. It not only illuminates a stunning mix of monarchical and democratic sovereignty but also gestures to what the sociologist of religion Robert N. Bellah called "civil religion in America."[15] The role of the ultimate sacrifice of the abstract citizen plays into the way knowledge, power, and law are distinct in democratic sovereignty while the invocation of God and the certainty of sacrifice as a product of God's divine will harken to premodern sovereignty. Reagan (1988) said:

> We could not forget them. Even if they were not our own, we could not forget them. For all time, they are what we can only aspire to be: giving, unselfish, the epitome of human love – to lay down one's life so that others might live. We think on their lives. We think on their final moments. In our mind's eye, we see young Americans in a European forest or on an Asian island or at sea or in aerial combat. And as life expired, we know that those who could have had last thoughts of us and of their love for us. As they thought of us then, so, too, we think of them now, with love, with devotion, and with faith: the certainty that what they died for was worthy of their sacrifice – faith, too, in God and in the Nation that has pledged itself to His work and to the dream of human freedom, and a nation, too, that today and always pledges itself to their eternal memory. Thank you. God bless you.

It is worth pausing here, before analysing the content of the speech itself, to recall how liberal democratic nations have a built-in right to call on citizens to sacrifice, and to ensure the democratic health and survival of the nation, they are obliged to answer the call. Recall that despite then prime minister Harper's own admittance that the meaning of sacrifice is ultimately contested (it is "difficult to measure"), he still nevertheless invoked it in concrete terms, as that which is integral to democratic sovereignty – drawing on both premodern and modern notions of sovereignty – of the image of sacrifice to make universal claims about the essential link between nationalism and militarization, as well as promote a distinct imagining of Canada's historical legacy and Canadian identity as something inherently vulnerable and that is being remade in relation to the historically unbearable weight of wartime sacrifices. I cited Harper's remarks that sacrifice could never be truly known or measured – in our secular world, who can legitimately stand in for the divine to represent sacrifice? Curiously, he nevertheless asserted that Canadians still feel the pervasively haunting yet unknowable presence of sacrifice.

Both Reagan and Harper shore up the symbolic capital of sacrifice to establish collective identity through war, but this is accomplished arbitrarily, because democratic sovereignty offers no body to represent the invisible/transcendental qualities of sacrifice, hence the continued references (some more explicit than others) to the divine world being the only source of the truth of sacrifice. Moreover, Reagan's speech opens and closes regarding the perseverance and indeed necessity of memory – "We could not forget them. Even if they were not our own, we could not forget them" and "the certainty that what they died for was worthy of their sacrifice – faith, too, in God and in the Nation that has pledged itself to His work and to the dream of human freedom, and a nation, too, that today and always pledges itself to their eternal memory." The implication here in the hierarchy of death (privileging those who committed the ultimate sacrifice) is that, to continue to be free, more death, and more sacrifices, may be warranted, for if future-potential wartime sacrifice fails, in both discourse and practice, then these deaths will be forgotten, and we will no longer be empowered to be free. Governing sacrifice implies both a contradiction between security and freedom and an interpolation of an empowered, sacrificing subject, whose sacrifice makes us free. The legacy of sacrifice, the memory of sacrifice past, attempts to sanction the continued liberal democratic call for continued sacrifice while producing and enabling citizens to answer that call.

To accomplish the making of an act as sacrificial or not, by rendering violence either "public" or "private," will depend on the goals and ideas tied to the application of violence. As Patricia Owens states, "there is no such thing as public or private violence. There is only violence that is *made* 'public' and violence that is *made* 'private'" (quoted in Eichler, 2015, p. 4). The same might be said for sacrifice – there is no such thing as purely sacrificial violence or purely non-sacrificial violence. What counts is how that violence draws on a certain symbolic regime of power; how it is regarded and represented as either sacrificial or non-sacrificial, and the varied sociopolitical effects that are wrought from each categorization.

Sacrificial violence is deeply interconnected with the social and political realms despite the prevailing ambiguity and religious connotations of the concept. While the Tomb of the Unknown Soldier is a product of democratic sovereignty, we ought to be curious about how the Tomb signifies remnants of premodern sovereignty's consolidation of knowledge, power, and law. This occurs through memorialization and memory practices, and results in the shoring up of sacrifice's place in a distinctly national politics of violence. How does sacrifice continue to make sense? It is worth here turning to the notion of civil religion.

Civil religion "at its best is the genuine apprehension of universal and transcendent religious reality as seen in or, one could almost say, as the experience of the American people" (Bellah, 1967, p. 12). This is exemplified in how Memorial Day, for example, "is a major event for the whole community involving a rededication to the martyred dead, to the spirit of sacrifice, and to the American vision" (p. 11). Bellah (1967) writes that "the American civil religion was never anticlerical or militantly secular. On the contrary, it borrowed selectively from the religious tradition in such a way that the average American saw no conflict between the two. In this way, civil religion was able to build up without any bitter struggle with the church powerful symbols of national solidarity and to mobilize deep levels of personal motivation for the attainment of national goals" (p. 13). The concept of sacrifice is one that survives and indeed thrives in civil religion.

Furthermore, Bellah (1967) writes that, in times of war, people must contend with mass deaths of soldiers (consider the Civil War, when as he claims the theme of sacrifice "was indelibly written into civil religion" and the great number of war dead required the establishment of several national cemeteries.) As such, sacrifice acts as a narrative bridge between private religion and politics, culminating in what Bellah describes as civil religion. This has two consequences: First, as Bellah notes, the theme of sacrifice serves to mobilize support for national goals (i.e., for our purposes, continued war-making, and the opportunity for moral growth, the unification of the collective, and redemption therein). Second, in the mobilization of support for national goals, the contradiction of sacrifice – its dual nature – is temporarily overcome; its ambiguity is reconciled in the realization of national goals. As Brighi and Cerella (2015, p. 12) show, the target of sacrifice – the scapegoat, in the case of foreign wars, the "uncivilized" – is an object of contempt (as cause of disorder) as well as veneration (for the re-establishment of order). This duality poses a contradiction that, as Asad (2007) writes, is often worked through but never resolved. The contradiction of sacrifice, how its dual nature is overcome, has much to do with modern liberalism's culture of war. As Asad states, "the cult of sacrifice, blood, and death that secular liberals find so repellent in pre-liberal Christianity is a part of the genealogy of modern liberalism itself, in which violence and tenderness go together. This is encountered in many places in our modern culture, not least in what is generally taken to be 'just' war" (p. 88). Furthermore,

> today, this contradiction is a part of a modern liberalism that has inherited and rephrased some of its basic values from medieval Christian tradition: on the one hand, there is the imperative to use any means necessary (including homicide and suicide) to defend the nation-state that constitutes one's worldly identity and defends one's health and security and, on the other, the obligation to revere all human life, to offer life in place of death to universal humanity; the first presupposes a capacity for ruthlessness, the second for kindness. The contradiction itself constitutes a particular kind of human subject whose functioning depends on the fact that the contradiction has to be continually worked through *without ever being resolved* [emphasis added]. (p. 88)

This invokes the notion of "dying to give life," which is reminiscent of Reagan's speech and Lincoln's address at Gettysburg, in which he famously states that soldiers give their lives so that the nation might live. The Civil War was one of the bloodiest wars of the nineteenth century; the loss of life was far greater than any previously suffered by Americans. It is no surprise, then, how sacrifice came to be a centre piece of American civil religion and wider consciousness. However, it did so during the emergence of *total warfare*. It was, to be crude, dead or disfigured bodies that made sacrifice a vehicle for civil religion, whose vocabulary can be translated and made palatable for even the non-religious, so easily finding a home in a public expression of secular religion, or American civil religion. Consider an additional definition of civil religion: "American civil religion is clearly an offshoot of the Judeo-Christian tradition, but it is not confined to conventional denominational categories. And while the concept of divine providence implicitly stands behind American civil religion, its character is, by definition, secular- it functions through such institutions as the branches of government, patriotic organizations ... and outlets of popular culture" (Angrosino, 2002, p. 241). American civil religion is therefore both religious and secular; somehow sacrifice, a deeply religious concept, must find a way to survive within political and ostensibly non-religious discourse.

As the American Civil War showed, as well as the establishment of national cemeteries and the Tomb of the Unknown Soldier, sacrifice *makes sense* in relation to wartime loss and suffering. If, however, that great loss of life no longer occurs due to political workarounds (privatization, fewer publicized troop deployments) in combination with technological solutions (drones and other distancing technologies), such that death in war becomes something to be calculated, as that which is to be avoided at all costs, then the role of sacrifice in civil religion risks losing its most compelling quality. Without bodies to attach itself to, the visual spectacle of sacrificial violence is not so readily translated to audiences, both religious and non-religious alike. The risk of neoliberal war (its detachment from bodies, and therefore society) is in fact the very same thing that justifies it. War risks becoming antisocial.

Summary

The aim of this chapter has been to understand sovereignty's relation to biopower in an era of governmentality and postmodern violence. Postmodern violence builds on the rationale of governing-at-a-distance, which is, in turn, emboldened by privatization. Therein, sacrifice is governed to temporarily reconcile the tensions between sovereign and biopolitical modes of power. As such, military violence today is distanced, privatized, and technologized, and therefore to some extent, and as will be discussed more in the next chapter, always-already disembodied. Disembodied war in combination with a notion of decentralized political authority means that there is no obvious ruler, no monarchical sovereign to make the sacred immediately apparent. I have argued that the increasing invisibility of the citizen-soldier implies that transcendental quality of sacrifice, as made significant through sovereign power, has quite literally *no body* to bear it and render it present. Still, the governing of sacrificial politics (how killing and dying are justified politically) transpires on and across the citizen-soldier body, as it is both a subject and an object of sacrificial violence. Moreover, that we accept the authority of democratic sovereignty does not imply that fragments of pre-modern sovereignty do not still circulate to give sacrifice weight in popular imaginings of nation-state violence. My analysis of the Tomb of the Unknown Soldier argued this point. Furthermore, as I suggest later, we can gesture to a "decline of the sacred" as is tied to the body of the citizen-soldier while still acknowledging modes of "re-sacralization."

Due to globalization and the neoliberal restructuring of capitalism, US-inspired militarized violence is becoming disassociated with the deployment of citizen-soldiers. As a result, violence is disconnected from the nation-state, that foundational basis which gives sacrifice meaning. Disembodied warfare, in essence, aims to overcome the requirement of bodies in a society that is both casualty-averse but endlessly preparing for and engaging in war. But absent bodies at risk on the frontlines, sovereign power struggles to attach itself to a suitable body to bring the myth of sacrifice into being. Subsequently, the sacrificial system is unhinged in response to this thinning of the citizen-soldier archetype such that the concealment of violence, and the concealment of the sacred (not its revelation) becomes the modus operandi of the democratic sovereign.

Furthermore, by treating, as I do later, the qualities of sacrifice (in relation to the People) as that which can be 'made up,' it follows that sacrifice can be unmade, reconfigured, or destabilized in relation to a changing style of violence. It is here, through three interrelated axes, where I next argue that sacrifice is displaced through the combined efforts of military privatization and drone warfare through the following mechanisms: publicity/privatization, surplus/rejection of surplus, and embodiment/disembodiment. While I treat these sites later in distinct sections, I suggest that they should be read holistically. While this text so far has largely been theory-driven, the following chapters include interview data to make better sense of the theoretical framework presented. My aim is to illustrate how drone warfare is a technological extension of the philosophy of military privatization. Going forward, my hope is that these empirical insights will help bring to life some of the key concepts and ideas discussed so far.

Chapter 4

Displacing Sacrifice

Violence flows out of the body or enters into it in order to create political value.
– Feldman (1991, p. 144)

Outsourcing War

Given that sacrifice is partly about the power to name and legitimize certain deaths as having public and/or national significance – including the power to recognize certain deaths or injuries as meaningful – the outsourcing or privatization of warfare, combined with the proliferation of distancing and disembodying technologies, such as drones, invites the question of what is at stake in the negation of the sacrificial violent encounter and its attendant, productive surplus. I define a productive surplus as the social and cultural capital that can be harnessed by the nation-state to merge war-making with collective, national identity. If war is a site for belonging, for social bonding of sorts, what would it mean then to reject the surplus that arises from a violent encounter, from wartime deaths or injuries? This rejection reveals a reconstituting of the *political value* that Feldman claims flows from and into the body in a violent encounter. Put differently, sacrifice's displacement is coterminous with the reconstitution of the politics of embodied violence in war. We can learn a lot about the changing status of war by thinking about how, and to what degree, sacrifice is silenced or summoned in national politics.

Recall that military sacrifice is not reducible to a single act but rather is indicative of a series of political conditions and discursive acts, which take shape within a particular and historically specific style of militarized violence. In this chapter, I argue that militarized violence obtains its social meaning partly through the recognition of the act of killing and/or being killed as sacrificial or not. This is accomplished through the demarcation of military violence as bearing either public or private significance. Since militarized violence is contingent upon the archetype of the citizen-soldier, and its implied sacrificial heritage and homage to sovereign power therein, any material attempt to overcome or reject the citizen-soldier archetype in the articulation of nation-state violence implies a parallel reconfiguration of its sacrificial content; likewise, a rejection of sacrifice is also a rejection of the citizen-soldier archetype. Therefore, the "decline of the sacred" can be linked to the decline of the mass army model of warfare, but this does not imply the end of sacrifice. It does imply, however, that we can no longer readily turn to sacrifice as that which links nation-state violence with citizen-soldiers as postmodern war is becoming less viable as a vehicle to ground a claim of national, collective identity. *Who cares*, you might now be asking. *Is this even a problem?* It is if you believe that for war to be justified, it must also be appropriately *social* with a direct path to the People who authorize it. Without sacrifice, a country might not feel they have "skin in the game," and this could, in turn, compromise the idea of democratic legitimacy in war. Outsourcing is one strategy among many that gives states the opportunity to override the requirement to have skin in the game.

What is outsourcing, and why does it matter? When I hire someone to provide landscaping services at my home, for example, I am outsourcing the job, which is a burden to me. I do this because I do not have time or energy to do it myself or because I lack the expertise to do a good job. Perhaps I simply cannot do it, for lack of time, desire, ability, or some combination of reasons. I do not have a green thumb, and as a result, I do not want to take responsibility for the foliage in my garden, including the possibility that I might make a mistake or harm something, as this would be bad for both me (I would feel bad) and the foliage in question (it might be irrevocably harmed). Hiring someone else who wants this task, who is more willing, competent, and able to take on any risk, makes sense. The outsourcing of war-related tasks works on a similar logic, albeit with dramatically different social, political, and ethical ramifications than found in caring for one's garden: States simply cannot, do not want to, or do not know how to conduct the globally expansive wars of today without the personnel and expertise offered by private military corporations (PMCs). Outsourcing violence, or outsourcing the means to give violence favourable conditions, implies that governments are in some ways burdened by war even as they are compelled to enact and sustain it. Outsourcing allows them to disavow some of these burdens, while granting plausible deniability when things go wrong. PMCs let states conduct war without drawing in a national population. Thus, states protect their own citizens, keeping them at a certain distance from the frontlines while easing the burden of war labour.

The ritualization of the sacrifices of citizen soldiers produces foundational narratives thereby enabling a communitarian effect. Unlike private contractors, citizen-soldiers' actions are consecrated within the boundaries of the nation and memorialized in a way that allows for both the production of shared collective memory and a projected future-oriented discourse of unification through shared national or ethnic destiny. As an archetype of citizenship, the citizen-soldier normalizes a belief that soldiers' actions in wartime reflected the highest echelon of sacrifice. Historically, military sacrifices have been publicly mourned and remembered.

Why might a state seek to avoid these collectivizing activities now or in the future? And why would this avoidance matter for both ordinary citizens and soldiers alike? What might we be trying to forget if we do not collectively aspire to embrace and acknowledge the violence we sign off on – *to remember it* (Zehfuss, 2003)? I am not making a case for excessive or violent nationalism. I am not advocating for some kind of return to an imagined past, a seemingly homogeneous, romantic, bloody, and glorified past. Instead, I am asking that *we be curious* about what is at stake, or what is lost, in the moments where war narratives and practices appear to disavow and deny sacrifice just as much as when they are encouraged or embraced. Must war have a strong nationalist or communitarian backbone, and what are the consequences one way or another? In the following, I discuss how the privatization, or outsourcing of war, is based on a philosophy of neoliberal rationality which produces two intertwined effects: (a) the rejection of the surplus in war (defined as that which can be harnessed by the nation-state for utilizing war as a site of collective identity) and (b) the desire to perfect a "new normal" of wartime violence, namely, *disembodied* violence. These are both socio-technical processes, meaning that they hinge on a co-constitution between society and technology. Together, these effects combine to create the foundation for drone warfare (the subject of the next chapter), which I claim finds its origins in the philosophy that underpins and justifies military and security privatization.

Publicity/Privatization

In the period of nation building, social relations were engaged within the nation-state's boundaries. However, globalized wars combined with military privatization threaten the citizen-soldier's status as a stable subject of sacrifice. Despite this, liberal governments still aim to control the communicative potentiality of sacrificial rhetoric. Part of the state's strategy for managing sacrifice is to privatize it and move the effects of the deaths of soldiers into the domain of the private family, thereby neutralizing them by placing it within a gendered framework of mourning (Butler, 2004; Taussig-Rubbo, 2009, p. 86). The relationship between citizenship and sacrifice has weakened with flexible citizenship, and the private contracting of soldiery captures a shift relative to the sacrificial logic that is bound up with the archetype of the citizen-soldier. Moreover, recall that "the problem of government at a distance in the 19th century led to an intense problematization of the ethical comportment of those who would govern, and the attempt to inculcate these ethical technologies through systems of training" (Rose, 1999, p. 148), resulting in the bureaucratization and incorporation of experts into the machinery of political government (p. 149). In securitizing the right to be free, private military corporations shore up expert knowledge necessary to engage in technologically advanced war in a way that also sustains agility and flexibility to meet the changing geopolitical and spatial demands of warfare. Technology is always changing, and presumably always improving; software is always updating, so militaries require constant and renewed flows of expert knowledge from outside their own organization.

The desire for expertise via flexible citizens and flexible production within the global economy penetrates thinking about military technology and soldiering – areas not excluded from market rationality and management expertise. As part of the revolution in military affairs (RMA) and its impetus to downsize, modernize, and professionalize the US military, Dalby (2008) states that emerging technologies enable soldiers to maintain a critical edge, to remain protected, competitive, and effective. The RMA

> is used loosely to refer both to technological innovations in weapon systems and in particular the most important changes wrought by computer technologies and communication systems. … Remote sensors and computer tracking of numerous targets supposedly allow sophisticated combat operations to outmaneuver foes and destroy opposition targets relatively easily with few casualties. This requires a reorganization of armed forces with large tank and infantry divisions broken into smaller units to be more flexible and capable of moving much faster over long distances. (Dalby, 2008, p. 1)

Note Dalby's emphasis on the importance of technology and casualty aversion – two practices that are theoretically drawn together with the implication being that the adoption and implementation of sophisticated technology greatly reduces casualties, which is a highly desirable outcome in war if not the most important measure of its success. Emerging technologies enable the covering of more ground – the expansion of human senses and capabilities – without physical boots on the ground, absent an embodied presence. As Blanchard (2011, p. 153) claims, drawing on John Arquilla, the perceived need to avoid casualties activated the RMA to stabilize defence spending in an era of financial austerity and to demonstrate the continuing usefulness of various military tools in an environment seemingly devoid of serious threats. Recall Leon Panetta's point that the US military will no longer be sized for large-scale operations. PMCs satisfy several goals to this end insofar as they provide the expertise necessary to facilitate the integration of emerging technologies and personnel to supplement or replace masses of troops through technologically intensive modes of warfare (Sassen, 2002, p. 280). Rather than an example of a newfound "virtuous" or "humane" style of war, the desire to remove humans from combat is reminiscent of war in the eighteenth century. In this period, war was a private, dynastic affair; it, too, did not significantly involve, target, draw in, or rely on ordinary civilians in any meaningful way.

Undoubtedly, private warfare has existed for centuries (Singer, 2005). Once a tool of monarchs who fought with private armies (Kinsey, 2006), the use of mercenaries diminished after the French Revolution and is now generally considered ethically, morally, and legally problematic for countries claiming to adhere to the laws of war. While PMCs trouble just war theory and the morality of war (Pattison, 2014b) private armies for hire make up most of human history insofar as how armies confronted one another on the battlefield. Yet, "our general assumption of warfare is that it is engaged by public militaries, fighting for the common cause. This is an idealization. Throughout history the participants in war were often for-profit private entities loyal to no one government" (Singer, 2005, p. 19). Indeed, the historical ebb and flow of the commercialization and bureaucratization of violence can be traced back to the emergence of the European state. Having once been a significant tool of monarchs attempting to seize and secure territory, again, the use of privateers diminished after the French Revolution when "wars became wars between nations, fought by citizens of those nations, as opposed to between monarchs with private armies" (Kinsey, 2006, p. 43). Suddenly, the story of the state placed the populace at its centre. The People, as individual embodiments of sovereign power, were now ideologically prepared to initiate and administer warfare themselves, on behalf of one another and against threatening outsiders. Liberal democracies today consider the People as central to the authorization and conduct of war.

However, just as the People came to be constructed as the legitimate source of state violence, such a concept, and its attendant beliefs and practices, can be gradually unmade too. The decline of the mass military model showed that the military was not exempt from the broader institutional transformations that marked industrial societies at this time of neoliberal reform: Like other institutions, the military too shifted from a labour-intensive to a capital-intensive organization, promoting casualty avoidance, real-time battlefield data, and agile weapons systems (Manigart, 2006). The decline of the mass army model complemented a restructuring that emphasized privatization, downsizing, and professionalization (Joachim & Schneiker 2012). Managers and technicians increasingly conduct war, as opposed to combat leaders (Moskos, 2000, p. 15). Importantly, the professionalization of the military was essential for realizing the technological goals of the RMA: "a policy agenda emphasizing the exploitation of technological advances to preserve and even improve the United States' long-term strategic position" (Shimko, 2010, p. 2). Without this professionalization, the incorporation of precision-guided weapons, for example, would not have materialized (Adamsky, 2010, pp. 59–61). But this downsizing and professionalization are also rooted in a contentious military history. As the following interview excerpt helps explain, the decline of the mass military model was partly an effect of how disciplinary issues intersected with budget cuts, changing mission parameters, and a growing distrust of the military in the post-Vietnam era:

> It's really about the transformation of military logistics, over the last few decades, particularly post-Vietnam, the US government has been trying to figure out how to, for reasons of cost, but also for reasons of morale, shrink the size of the US Army. Kids don't want to go to war. They don't understand the need for it, so when you send people to war using a draft you're gonna get some very reluctant people who are maybe unruly; a lot of people deserted, a lot of people messed up with drugs, a lot of people even shot their officers, so you had a real problem in trying to get people to the war. (PC, interview)

Furthermore, military veteran AS adds, regarding the Vietnam war:

> It wasn't for God and country – "let's do our time, stay alive and get out of here." A number of guys I served with did that kind of thing and what

> I picked up over the years from them was a general feeling that Vietnam was, you know, a mistake but that basically the military was used for political purposes and by people that had really no interest in the men and women serving in the military, that they were just a tool, and of course with the loss of life and the element of defeat, there was just a sense that we're not gonna let this happen again. (Interview)

Following the war in Vietnam, the decline of the mass army model coincided with casualty aversion and the entrenchment of economic liberalization policies. Citizenship became reoriented away from public or nationalist ideals towards private, individualist, and consumer-based notions of subjectivity and knowledge. This has occurred alongside the development of the securitization of citizenship as a chief concern for the state. By securitizing citizenship, "technologies of governing are employed that displace the governing of populations from authorities and sites traditionally located in the state to other sites and actors such as private companies, international organizations, and even individuals and their own self-government" (Rygiel, 2008, p. 211). Hence, "citizenship as government is not just internationalized but also *privatized* and *individualized*" (Rygiel, 2008, p. 211). PMCs reflect the need for patience over wrath; rather than unconstrained and even unreliable mercenaries, they are the product of rational considerations and a political culture that includes hyper-individualism, techno-centrism, and a growing set of security problems. After 9/11, PMCs were hailed as necessary actors in the successful implementation of the goals of the RMA, but how states would acknowledge their presence in the battlefield was less clear. Still, "simply put, one could not tell the story of the Iraq war without any discussion of PMCs" (Singer, 2004, p. 6). Like my problems with gardening stated earlier, states simply cannot imagine contemporary governing without outsourcing, and war is not exempt from this – neoliberal rationality pervades all dimensions of social and political life, including war.

PMCs are the effect of the widening of the meaning of security such that the boundaries of the nation-state were no longer capable of effectively administrating security as dwindling Euro-American state militaries were thought to be ineffective in new wars (Duffield, 2001). Thus, there was an unprecedented intervention of PMCs into Iraq following the American-led 2003 invasion, which was justified through the false assertion that Iraq was harbouring weapons of mass destruction, among other claims that falsely linked Iraq with the terrorist attacks of 9/11. Simultaneously, the American government outsourced many of their operations to PMCs, rendering the war effort unprecedentedly reliant on non-state, corporate actors (Singer, 2004). In 2007, for example, there were more private contractors than US soldiers on the ground in that country (Hartung & Pemberton, 2008). At the height of the war, private contractors outnumbered US troops, and some have estimated that there were as many as 50,000 armed security personnel (Singer, 2004, 2005). Consider the following interview excerpt regarding military privatization. "V.H." served during the Cold War and joined the National Guard after 9/11. They did a tour in Iraq in 2004–2005. In 2011, they deployed to Afghanistan as a private contractor:

> VH: I met a couple guys [contractors] early on in Iraq, and they were crazy. And one guy talked, told horrible stories, about what he had done. Were they true? I have no idea, but it was really disturbing stuff. And the other guy, I don't know what he looked like … he looked like a cartoon character commander type guy. Weapons all over him, bald head … who knows. One theory that I've heard, the Black Water guys work for the state department for the most part, not the military, and they were used to do things that the American military couldn't get away with.
>
> BB: Such as?
>
> VH: Assassinate people, and do stuff like that, to do stuff to keep the military's hands clean. It's like the CIA. Is that true? Who knows, but it makes a certain amount of sense. Those guys were above the law compared to the army units. … The military hates the contractors, and they all want to be one when they get out. Because the contractors make a lot of money, a lot more money than the soldier does. And so they totally resent it. And this started happening in the 1980s. What I used to do in the Army, contractors do now. I guess part of it is that the military isn't big enough. We would have to have the draft to have that many people. … I went to Afghanistan as a contractor for financial reasons. I got an offer I couldn't refuse. I was unemployed; my house was about to get foreclosed on. They said, "Can you be in Fayetteville in two weeks for training?" "How much? OK I'll be there!" And then of course, contractors don't carry weapons, but I saw more combat in Afghanistan as a contractor than I did in Iraq. I saw a lot of indirect combat, but anyway. I struggle with this all the time. I write poetry about it. (Interview)

As VH attests, PMCs do the stuff needed to keep the military's hands clean, the stuff they perceive they cannot do, be it for logistical, ethical, or political reasons. They offer services in information technology, coordination and support, military training, combat and security (Alexandra et al., Baker, & Caparini, 2008), supplementing weakening and shrinking state armies. They also satisfy neoliberal socio-political and economic desires for techno-rational, geopolitical, and scientific mastery of space and time. As such, they are now conceptualized as integral to how states imagine and manage security threats. PMCs thus reflect, respond to, and further blur the distinctions between civilian and combatant, war, and peace (Kinsey, 2006).

Critically, wars are no longer fought on battlefields between opposing armies wearing uniforms. Instead, the battlefield is everywhere: cities, towns, and countryside (Kinsey, 2006, p. 52). Battlefields are virtually constructed and simulated (Der Derian, 2001), while security threats are conceptualized as having transnational origins and effects. In the post–Cold War period, "new" threats, enemies, and transnational security issues emerged alongside expanding battlefields, such as the drug trade, environmental degradation, poverty, and immigration (Moskos, 2000). PMCs are ostensibly a permanent fixture in the security industry because of these seemingly new threats and their ability to fill the gap that asymmetrical, ethnic-based, humanitarian wars create. Privatized warfare is a site where economic and military objectives intersect, and this alignment cannot be separated from the dynamic social meaning attributed to war, combat, and sacrificial violence. The significant incorporation of private military contractors into the application of military violence shows how the discursive and material qualities of sacrifice are contested in relation to these politics. PMCs benefit from neoliberal governmentality's reliance on expert knowledge to develop truths about, and solutions to, security threats. In this way, just like any other business, they can manage their self-presentation due to their ability to self-cultivate images of legitimacy, superior expertise, and as adaptable to changing security contexts (Joachim & Schneiker, 2012).

However, in addition to the authority granted to PMCs and their affiliated personnel insofar as their ability to generate, curate, and provide solutions to security threats, thereby confirming their self-ascribed credibility, "neoliberal governmentality tends to 'de-politicize' security as public debate narrowly focuses on the technicalities and costs of military solutions, while alternative political options ... become marginalized" (Leander & van Munster, 2007, p. 201). PMCs' embrace of self-governance is simultaneously a function of their participation in the de-politicization of war and security. Duffield (2001) similarly explores how war and corporatism (which capitalizes on the idea that security threats are prolific and unending) are co-constitutive. I argue that the privatization of security and military operations has a threefold effect: the aesthetic sanitization of war, the disruption of the sacrificial system characteristic of liberal democratic nation-states, and, finally, an enabling of privatization's technological counterpart – the cultivation and application of drone warfare, the subject of the next chapter.

Because PMCs are deployed secretly and contain no internal requirements to publicize their operations, an ideological impetus to control and sanitize the publicization and/or privatization of death, violence, and suffering, for largely American and Western audiences, is provided as both just and rational. As a criticism, feminist geopolitics show how Iraqi civilian deaths, for example, are marginalized and subordinated in relation to American deaths – deaths regarded as politically qualified (Masters, 2007). Hyndman (2008, p. 199) claims that, geopolitically, the question of what kind of suffering is privileged in wartime is related to the questions, "Who counts?" and "Who cares?" "The 'fatality metrics' of body counts is deeply lopsided in this context: victimhood is commodified and patriotism publicized for soldiers making 'the ultimate sacrifice,' while Iraqi deaths are framed as 'the price that must be paid' for introducing 'freedom and justice'" (Hyndman, 2008, p. 199). Butler (2004) further explains this dehumanization process as it relates to racial discourses and the production of Agamben's concept of bare life, which he describes as receiving a kind of (non-)violence: "Violence against those who are already not quite living, that is, living in a state of suspension between life and death, leaves a mark that is no mark" (p. 36). The corporatization of warfare is an effect of a political need to dictate and control the terms and perceptions of American suffering, rendering the abstract "American-ness" as that which ought to be protected. This flexibility of control, signified by the trend to plainly make the violence of war a private matter, is solidified through the incorporation of PMCs personnel – subjects who cannot sacrifice for a nation and are therefore expendable through their very subjection to the flexibility of control offered through neoliberal rationality and privatized models of warfare.

Consider the case of Texan Jamie Leigh Jones as an example of how military privatization quite literally *makes violence private*. Jones was drugged and gang-raped in her mixed sleeping quarters while working in Iraq in technical support for the company KBR. Jones endured fifteen months of arbitration[1] before she could file a civil suit. This is because she unknowingly signed a contract stating that any "workplace dispute" would be settled in private arbitration rather than a public court. Even if employees do read the fine print, they often do not realize they are waiving their right to a jury trial (Summers, 2004). Arbitration meant that KBR had the right to unilaterally hire the mediator, all proceedings were secret, and Jones had no right to an appeal. Arbitration requires no public record or transcript of testimonies or proceedings.

There is no formal mechanism for courts to penetrate the reasons for the arbitrator's decision; arbitrators, unlike judges, do not have to write opinions stating their findings and thus are not held accountable to their peers (Summers, 2004, p. 708). Arbitration is supposedly better for workers and more economically efficient; KBR claims that this process helps protect the identity of victims. Indeed, arbitration is better *for KBR* because juries are more favourable to employees than arbitrators and tend to award larger compensatory and punitive damages (Summers, 2004, p. 693). To be sure, the protection of identity is important to KBR but for different reasons than they would offer. This is because the protection that KBR is committed to is also a part of a desire to privatize and individualize experiences of what amounts to, in instances of gendered violence and harassment at the very least, *collective* acts of violence. This is not to negate the individual responsibility of perpetrators of violence. Rather, the fact that arbitration prevents class-action lawsuits suggests that corporate employers benefit financially and in terms of protecting their reputation when they can isolate victims and eliminate the opportunity for publicity of any kind, thereby preventing the formation and solidification of a collective sentimentality that could be harnessed against the employer. At the time of writing, KBR has won 80 per cent of its arbitration hearings, perhaps because arbitration allows the employer to endlessly delay proceedings, quickly escalating costs and thus financially and emotionally wearing down employees (Summers, 2004, p. 714).

The denial of a date in a US court, and the binding arbitration Jones inadvertently agreed to, privatized her suffering, rendering it inaudible and non-political. Forced arbitration is part of how the law normalizes militarized violence and gives cover to PMCs' through a lack of transparency and accountability. PMCs have been accused of participating in forced prostitution and sex trafficking of women and children (Lasky, 2006), as well as inflicting sexual violence against Iraqis in detention centres (Kivlan, 2008). Yet, according to Singer (2004), not one has been prosecuted or punished for a single crime, sexual or otherwise. Thus, "we can only conclude that with PMCs in Iraq we have somehow stumbled upon the perfect village, in the midst of a war zone, where human nature has somehow been overcome, unlike in the most bucolic villages. Or, we have a clear combination of an absence of law and political will" (p. 13). Senator Al Franken harnessed Jones's case to advance his first major piece of legislation, an amendment to the Defense Appropriation Bill, which prevents the military from contracting with companies that force private arbitration on sexual assault or harassment claimants (Mencimer, 2011). Jones testified at a judiciary committee meeting in 2009, emphasizing her constitutional right to have her day in court. This trend towards arbitration (although it is now illegal for the government to contract with companies that force arbitration in instances of sexual violence) fits neatly within a broader political-economic mandate of PMCs and, by extension, labour relations more generally: In an era characterized by neoliberal global economic imperatives, militaries are not exempt from wider social, political, and economic reconfiguration of the public and private spheres, which sees the de-politicization of war, security, and violence.

When PMCs are implicated in discourses and practices of sexual violence, be it against civilians, racialized "others," or co-workers, the mechanisms that allow the corporate bodies to privatize knowledge of sexual violence effectively privatizes the harm and sever(s) the democratic impulse, which has come to represent the relationship between citizens and states. Wartime sexual violence is a global problem that goes beyond the terrain of the nation-state, and it still does not register as an issue deserving sufficient treatment by political elites. PMCs conceal the problem of (sexual)[2] violence, rendering its effects invisible. Meanwhile, PMC personnel cannot sacrifice for the nation; their deaths are not publicized, they carry no flag, nor do they function to unite a national community towards a common goal.

The RMA gives further credibility to PMCs in such a way that allows them to render their practices non-political (and thus non-sacrificial) through its claims to be practising healthy neoliberal economics. The RMA made technological advancement a key component of casualty avoidance and contains a complex set of practices that emerged, in part, within the dictates of global capitalism. Flexible citizenship was an integral model that freed both military and global economic objectives to facilitate the compression of time and space in mobile and transnational sites of capitalist production, reduce human casualties during war, and allow for efficient strategies of global military, political and economic dominance. It includes "technological mastery, omnipotent surveillance, real-time 'situational awareness,' and speed-of-light digital interactions" (Graham, 2008, p. 37). These developments reflect a profound desire to overcome the limits of the body of the citizen-soldier. This strategy, which included "a completely automated weapons system devoid of human involvement" (Graham, 2008, p. 51) is packaged in the language of economic efficiency and the protection of life, because it seeks to substitute bodies with technology. The increasing reliance on technical reason is not a coincidence. It is inextricable from privatization and flexible citizenship.

It should come as no surprise then that the RMA borrows ideological language by referencing "trade liberalization," "economic reform," and "free markets" – classic hallmark euphemisms of neoliberal-style capitalist accumulation (Parenti, 2007). PMCs are thus a product of the neoliberal restructuring of capitalism. They reflect a new phase of militarization in the post–Cold War era as well as the synthesis of the interaction between the sovereign power to kill and the RMA's biopolitical rationale of enhancing life. Their inclusion, therefore, aids in the state's ability to disavow sacrifice, thus expanding and legitimizing the mandate of extra-sacrificial wars, wherein bodies are both being killed and killing but without the earlier guarantee, promised by sacrificial cults, that these actions would be made meaningful. Importantly, PMCs provide the sociopolitical foundation for drone warfare. Put differently, drones are the technological extension of the philosophy of military and security privatization. Yet, so far, much of the discussion about PMCs and their inclusion in practices related to drone warfare has focused on legal (rather than ethical or political) questions of accountability and transparency particularly as it relates to targeting and related decision-making.

Laura Dickinson (2015) explores the risks in outsourcing aspects of the US drone program to contractors. Commenting on a report by the Bureau of Investigation about private contractors' involvement in military drone operations, which have an especially significant presence in drone imagery analysis, Dickinson asserts that, within state-sanctioned violence – killing, or so-called inherently governmental functions – it is not always entirely clear cut who is doing what. Despite laws preventing contractors from engaging directly in combat, interviewee VH talked about seeing more combat as a contractor than a soldier. In interrogations and decision-making regarding who is to be selected for targeting, Dickson writes that the lines between defence and offence, or what counts as targeting, are difficult to draw, because private military personnel confuse authority structures present within a military organization (Wong, 2006). The legal and moral responsibility for contractors on the part of military officials is unknown. Additionally, there is no formal arrangement for intelligence sharing and this exacerbates danger and friendly fire incidents. Contractors also challenge unit cohesion. Whereas soldiers historically report that they fight primarily for each other (Elshtain, 2019; Woodward, 2008) under the flag of their country, or for ideological commitments relating to freedom, liberty, and democracy (Wong, 2006), private contractors are not motivated by the same ideals. PMC personnel, although typically having more experience than uniformed personnel, are motivated by professional and economic gain. They are often former military who, given their already established credibility, engage in private contracting to earn additional income, gain skills, and remain relevant. The philosophy of military privatization prioritizes armed protection for individuals, places, and things; it does not consider respect for law and public safety as the primary aim (Duffield, 2001).

The military–industrial complex illustrates the logic underpinning military privatization. There is now a vast and permanent network of contracts, flows of money, and lobbying between people, corporations, institutions, defence/military contractors, and in the United States – the Pentagon, the executive branch, and Congress. This is evidenced by a revolving door between the military and political spheres. It is reasonable to expect that when a service member retires from the military, they will go into the private sector, where the demand for that kind of expert knowledge is widespread and the pay is significantly higher as compared to the services. The military–industrial complex also symbolizes the culmination of a new relationship between public and private realms beginning in the nineteenth century. Modern warfare was becoming so complex – such that weapons became *systems*, compromising not just a weapon but also a platform and means of command and control (Kaldor, 1977) – that it required large components of industry to be devoted to research and development to produce, sustain, and improve rapidly developing military technologies.

This need for specialized knowledge meant that private companies would provide expert knowledge, serving as key players in the expansion of military bases and in support of proxy wars. During the Cold War, the ideology of nuclearism focused on technology platforms such as jet fighters and submarines. Smaller armies were preferred because air-delivered weapons could theoretically replace ground forces; therefore, the overall number of soldiers declined. Scientific and engineering labour became significantly more important than manufacturing labour and new forms of secrecy emerged to protect this rapidly developing technical knowledge. Yet the military–industrial complex hides the interconnections between structural and physical violence; violence is not an exceptional practice but is a product of the widening gap between global rich and poor, geopolitical instability, global competition, environmental catastrophes, new forms of imperialism, and old/new racisms. The notion that "we" (the "civilized" West) engage in war only when provoked, or in defence, is a myth that continues to circulate in favour of consolidating or resuscitating sovereign power.

The post–Cold War period extended military restructuring (as many institutions did in the 1990s) in line with common sense business practices: this included downsizing through outsourcing, (training militaries of other countries to do proxy work for US interests – this has the added convenience of being able to maintain plausible deniability when human rights abuses occur – although there are many reasons for issuing proxy work) and, of course, privatization and outsourcing, practices that benefitted from humanitarian wars, or wars of choice: New war-making doctrines expanded to other areas once governed by civilians such as famine relief, disaster relief, and evacuation operations, often encouraged by the ongoing escalation of civilian wars and the slow and often insignificant responses from the United Nations.

Moreover, in the post–Cold War era, where these types of proxy wars are the norm, these political and economic nuances of violence as perpetuated by the ideals of the American empire have a way of being subsumed by the twin banners of "democracy" and "freedom." "War" conjures up a now outdated rationale that it is ethical, legal, and bound by internationally agreed-on conventions as far as accountability and transparency are championed by states that wage war with other states. Indeed, asymmetrical warfare appears to be the default way of strategizing violence – how to contain it and prevent it from spreading elsewhere. Even the United States, which has been involved in several counterinsurgencies since 9/11, acknowledged these difficulties. The 2009 *US Government Counterinsurgency (COIN) Guide* (US COIN Initiative, 2009) indicates that insurgency is not always conducted by a single group with a centralized command structure, but may involve a complex matrix of actors with various aims. They may also be loosely connected in dynamic and non-hierarchical networks. Yet, the "enemy" is often portrayed to be homogeneous in ideological orientation.

As discussed, military/security privatization exploded with America's 2003 invasion of Iraq. Foreign policy at this time, as summarized by both Isabelle V. Barker and the 2002 National Security Strategy, was defined as the United States extending the benefits of freedom across the globe. "We will actively work to bring the hope of democracy, development, free markets and free trade to every corner of the world" (US White House, quoted in Barker, 2015, p. 84). Christine Delphy (2003) has termed this the imperialist White Man's Burden, a missionary's ethos that was later picked up by American political elites in the latter half of the twentieth century, specifically expressed within the discourses deployed prior to the most recent invasion of Iraq. For example, the Bush administration famously claimed that American troops would be greeted as liberators upon their arrival in Iraq. Furthermore, the Bush administration, through its neo-imperial framework and its use of "embedded feminism,"[3] essentially re-invoked past colonial constructs that have traditionally classified non-Western cultures as backward, lacking agency, and in need of Western intervention. As Delphy (2003) argues, the words have changed, but it is not difficult to recognize behind this new phrase, "the right to intervene," the same old white man's burden, still as lethal, for it incorporates the missionary's paradox: "We will save their souls [their freedom] even if we have to kill them to do it" (p. 344). During the Skyes–Picot Agreement of 1916, which divided Arab territories among the British and the French, General Maude, upon his entry into Baghdad, had unequivocally declared that the British had arrived as liberators, not conquerors (Abdullah, 2003). The missionary's paradox still circulates within American foreign policy under the guise of orientalist protection myths (Said, 1979). Moreover, antiterrorism discourses blends with neoliberal economics to enable governing through distance, patience, and calculation – but not through explicit control. This reflects the imperatives of governmentality that are integral to the logic of military/security privatization.

To govern in this manner means that "over half a million U.S. troops, spies, contractors, dependents, and others are now stationed on some 737 military bases located in more than 130 countries, according to official Pentagon inventories" (Johnson, 2008, p. 21). These contractors operating within PMCs come from all over the world to participate in some of the most lucrative work available in their profession. They reflect various socio-economic backgrounds, personal and professional histories, roles, and intentions. Yet, Barker (2009) claims, there is a strategic paradox at the heart of today's *Pax Americana:* "The only way the United States can support an empire of military bases [to accomplish the aforementioned National Security Strategy] with a trimmed-down force comprised of all volunteers is through outsourcing services" (p. 85). Military privatization reflects neoliberal ideals because it treats security like a commodity which operates within the confines of business ethics, something to be packaged and sold in the marketplace and consumed as a service to enhance efficiency and reduce costs.

However, in practice, this claim to economic efficiency did not translate into savings, or efficiency, in Iraq. Upon seizing Iraq in April 2004, occupation authorities fired all 400,000 soldiers in Saddam Hussein's army with the hopes of training 40,000 new soldiers across 27 battalions (Chatterjee, 2004, p. 125). This goal was missed by a large margin, yet this is where American-based corporations profit *regardless of whether the intended target is met or not.* For example, the Iraqi army was trained by US-based Company Vinnell, which was awarded US$48 million for the job while they paid soldiers a mere $70 per month. One major source of tension was the "forced integration of ethnic Arabs and Kurds, traditional enemies. ... American planners thought they could create a model for the country's diversity ... from the first day this was a nonstarter, because military training had to be translated from English to Arabic and then to Kurdish" (p. 126). Within the first few weeks, 100 Kurds quit – they also complained that their weapons malfunctioned. The US military eventually fired Vinnell for contributing to ethnic tensions among soldiers (p. 126). Despite Vinnell's failed contract, the Virginia-based company has a long history of providing services to the American military, from Guam in the 1950s to the Korean War (p. 126). Privatization has occurred in other areas besides those concerned with policing or combat. For example, Halliburton managed to secure US$3.9 billion in contracts related to transporting and maintaining equipment from the military in 2003, although it is estimated that contracts for Halliburton were ultimately worth as much as US$13 billion. Halliburton has been repeatedly accused of significant overbilling, attempts to beat the competitive bidding clause, and deliberate, widespread inefficiency (Chatterjee, 2004). Moreover, Bechtel, one of the world's largest engineering-construction firms, was awarded a US$2.8 billion contract to refurbish Iraq's sewage, water, and school systems. However, the company has a long history of botched reconstruction jobs and of creating economic hardships for local people once they vacate the county in question. According to one source, "Bechtel and privatization go hand in hand. When Bechtel comes to your town, you can expect costs to soar and accountability and local control to evaporate" (Beck, quoted in Chatterjee, 2004, p. 65). Bechtel recently brought a US$25 million lawsuit against Bolivia for cancelling a contract to manage the Cochabamba water system. Under Bechtel, the water rates for locals skyrocketed (p. 65). Within days of the fall of Baghdad, thousands of local and expatriate contractors, working for multinational corporations, were hired to reconstruct the country, and install democracy at a profit that most assumed would be paid for from Iraq's vast oil wealth (p. 13).

Moving beyond the immediate post-9/11 era, what can be said about military/security privatization in and beyond Iraq as it transpired between the Bush and Obama administrations? Where Bush's mandate included a costly nation-building principle (however misguided) Obama's reflected a similar pattern – indeed an extension of the philosophy of privatization – albeit with a so-called light footprint approach, which has relied upon thousands of Americans paid to fight and die in the shadows (Zenko, 2016). Under Obama, more private military personnel died in Iraq and Afghanistan than all the

US troops deployed to those countries (Zenko, 2016). Despite this, the Department of Defense (DoD) still does not have a reliable system for tracking contractor personnel killed or wounded, nor does it have a reliable way of knowing the citizenship of the deceased. Drawing on data published by the US DoD and the US Department of Labor, Micah Zenko (2016) cites one estimate that between 2001 and 2010, 32 per cent were citizens, while 68 per cent were non-Americans hired by US or non-US firms that had won a military contract. He also reveals that private contractors outnumber American troops, constituting two of the most contract-dependent wars in US history: At one stage in the war, there were roughly 28,626 contractors in Afghanistan, compared to 9,800 US troops. In Iraq, 7,773 contractors support 4,087 US troops (these figures do not include those who support the CIA or other intelligence communities; Zenko, 2016). Further, the presence of contractors encourages mission creep, since contractors do not count as "boots on the ground" and therefore do not count against troop-level caps; "as a result, the government can put more people on the ground than it reports to the American people, encouraging mission creep and rendering contractors virtually invisible" (McFate, 2016). An exception to this invisibility was the widely publicized incident at Nisour Square, Baghdad, when a group of Blackwater employees killed seventeen Iraqi civilians and injured twenty-four. Four guards were convicted in a US court. Blackwater changed its name to XE Services in 2009 and then again to Academi in 2011 (curiously just one letter short of *academic*, signalling its own self-proclaimed expert knowledge).

Indeed, the character of war today, its unprecedented complexity and scope across multiple domains, zones, enclaves, across time and space, shows how contemporary war is driven towards perpetual temporal and spatial expansiveness, even while certain elements pertaining to the ethical status of bodies in war *remain invisible.* This "growth market" gives the military–industrial complex endless streams of possible revenue. PMCs, like all profit-driven enterprises, will seek to exploit and benefit from this where possible, and the neoliberal restructuring of capitalism has positioned states, and casualty-averse, budget-strapped militaries, to gravitate towards outsourcing as solutions to problems of governance today. In this chapter so far, I have established military outsourcing as a new normal in warfare. But what are the social, political, and ethical effects? As the next section discusses, the privatization of war has two important effects when it comes to the governing of military sacrifice: the rejection of the surplus that historically emerges in parallel with citizen-soldiers' deaths in war and the increasing use of disembodied methods of war. I take up the first next.

Surplus/Rejection of Surplus

One effect of military privatization for thinking about sacrifice's displacement is in relation to the idea of a surplus – a productive surplus is that which can be harnessed by the nation-state for utilizing war as a site of collective identity. I use the term *surplus* to refer to the productive political value (the excess symbolic capital) that remains following the deaths of citizen-soldiers. I argue that, traditionally, this surplus has been productively harnessed by the state in its post-war remaking and re-imagining of itself and its history and future potential but that, in our contemporary moment, the inclusion of private military contractors reflects a rejection of this surplus political value. Georges Bataille, Karl Marx, and others have deployed this term. Bataille uses the term to outline his notion of a "general economy" (Bataille & Hurley, 1989). Writing against the classical political economy, Bataille challenges the notion of scarcity as being the guiding economic principle and instead posits surplus (growth, the accumulation of wealth) as the economy's defining feature. For Bataille, this wealth must be deliberately wasted. Grand displays of human power (e.g., in the form of wars and human sacrifices) reflect this destruction of surplus energy. In contrast, I use the term to showcase changing meanings of violence as ascribed to citizen-soldiers.

Taking the body as a site of transaction, consider Elshtain's (1993) analysis. Elshtain links sovereignty, masculine violence, and war together, by claiming "the state's proclamation of its own sovereignty is not enough: that sovereignty must be recognized. War is the means to attain recognition" (p. 162). As such, sovereignty, much like national borders, does not exist a priori. Instead, sovereignty is brought into being in relation to sacrificial violence, wherein soldiers act as *embodied combatants* and therefore as sacrificing bodies. For sovereignty to be realized, it must be made visible and recognized. However, this assumes the presence of embodied combatants in war. In the avoidance of violent encounters in the form of an exchange, characterized by the absence of a relation between embodied combatants, sacrifice is destabilized through disembodiment as an effect of an invisible, "unrecognizable" sovereign power.

In post war periods, when nations collectively reflect on the value or meaning of violence, fallen soldiers remain, in some symbolic sense, the property of the state. The nation reclaims this loss as a positive surplus in the remaking of the state. As the epigraph by Feldman illuminates, violence has a political value, or surplus, as it flows in and out of bodies. But drones generate a type of violence that does not *flow in between* as an exchange but, instead, unilaterally identifies, stalks, fixates, targets, and erases human life from a distance. That surplus is not recovered by the nation-state, whose audience is not directly engaged in a highly calculated yet simultaneously disengaged act of killing. Sacrifice historically has allowed nation-states to organize collective violence by using the symbolic excess that resides from a sacrifice to assert its authority (Edkins, 2003, p. 95). For example, when sacrifice is invoked alongside citizenship, belonging, and wartime remembering, it provides the grounds to effectively recast national collective consciousness in violent terms thereby renationalizing state power.

When a member of the US military dies in battle, for example, this is considered the grandest of sacrificial acts for the nation (Pearce, 2010), and the nation consumes this death and recirculates the spiritual residue of the fallen into the bloodstream of the body politic. Sacrifice traditionally implies the power to name and legitimize certain deaths as having public significance, recognize certain deaths as legitimately lost, and rank those deaths according to a schematic of who is worthy of recognition. Citizen-soldiers pay "the ultimate price" and are therefore, alongside their families, ranked highly in the national schematic of suffering and loss. But PMCs operate in the shadows. The invisibility of contractors and subsequently their deaths and injuries, the invisibility and refusal to make public the precise numbers and nationhood of those deployed, the invisibility of the light footprint's paradoxically expansive and heavy underbelly – together signify the rejection of the surplus traditionally inherent to the meaning of death and dying on the battlefield. In contrast, in the acceptance of the surplus, therein was the opportunity to turn potentially meaningless violence towards that of the ultimate sacrifice for the greater good of the nation-state. The presence of contractors also reveals what is absent: the embodied, sacrificial will of the citizen-soldier and, as a result, the People, whose name wars are fought on behalf of. The privatization of war drives a wedge between society and the violence committed on its behalf. Under these circumstances, it is no wonder how distancing technologies, such as drones, would be applied in a way that both confirms and amplifies this material and ideological distance. Disembodied war, which drones support, is really about the desire for *bloodless* war. This, in turn, alters the symbolism of sacrifice as it relates to war.

Let us circle back to the previous discussion of military/security privatization to both summarize the main points while further elaborating on the meaning of surplus and what I claim is the current rejection of this surplus. If as I described a positive surplus is that symbolic capital that can be effectively harnessed subsequent to wartime death, injury, or suffering for the purpose of utilizing war as a collective site of identity, then military/security privatization is an unambiguous rejection of that surplus: The sacrificial excess has no one to capitalize on it, and it has nowhere – no clear national audience and corresponding sovereign figure – to be absorbed, interpreted, and recirculated anew within the body politic. In the dialects of visibility and invisibility, privatization has a dual role: Its expert knowledge contributes to the rendering of the optics of the battlefield into the scope of the knowable and visible. Meanwhile, the bodies deployed under the mandate of privatization are made invisible, insofar as private military companies do not report to Congress; the US government does not publicize their inclusion in conflict zones, which outright deflects attention away from the topic and, as discussed later, provides no reliable method of ascertaining knowledge about contractor deaths and injuries.

Embodiment/Disembodiment

I have suggested that the content and meaning of sacrifice, especially in relation to globalized wars, is deeply political and cannot be theorized universally but, rather, only in historically specific contexts of war-making. Our current moment, evidenced by postmodern, techno-fetishistic war, implies not the reality of bloodless war, or the end of death or dying for one's nation, but a clear desire for this based on the biopolitical imperative to enhance life. Bloodless war is not a reality but exists on the horizon towards the destination of "humane" war. It is evidenced by a trajectory of increasing unevenness and invisibility of violence as a by-product of sovereign power's departure from the body of the citizen-soldier. War is made invisible to national audiences and therefore takes on an anti-political quality, thereby becoming reliant on widening the gap between citizens and the control/application of violence. The imagery of military sacrifice – as being tied to the vitality of the nation-state and the embodiment of sovereign presence, bloodletting, transcendence, and moral achievement through the destruction of the finite body and ultimately injury or death – now gains content only in superficial moments, being mobilized when nations attempt to periodically militarize citizenship through premodern iterations of sovereign power.

Nation-states develop with an interest precisely in securing territory, acquiring resources to secure war-making technologies, regulating populations, and categorizing individuals based on ethnic markers and citizenship. State policies like citizenship and border security not only secure the nation-state from outside interference, but they also solidify an ideological allegiance or loyalty to the state as seen in supposedly patriotic acts, such as defending, dying, or killing for one's homeland. The privatization of war, in contrast, demonstrates how the cultivation of this kind of loyalty is no longer an explicit priority of the state. This is because military outsourcing reshapes the very need for sacrifice in a period marked by supra- and non-state entities: States can now surpass borders and harness the appropriate forms of technology to justifiably incorporate non-humans and non-citizens into the practice of warfare vis-à-vis the interrelated logic of free-market capitalism, the decline of the mass army model, and flexible citizenship, promoting various forms of governing at a distance. In addition, or alternatively, they can also integrate advanced technology to generate "mass," absent sufficient human-centred capability.

As a tactic of governing at a distance, drone surveillance and drone strikes reveal a clear avoidance of a sacrificial *scene* – the familiar optics and players integral to battlefields of the twentieth century. In drone warfare, there is no formal embodied performance of a violent exchange in the name of the sovereign and/or nation, but only air power stripped down to its essence. Precision bombing (absent the pilot) is underpinned by detached, mundane calculations aimed at saving American life (Martin & Sasser, 2010). It is no surprise then that during the peak years of the war on terror, photographs of coffins draped in flags containing dead soldiers no longer circulated for public consumption as they once did during the Vietnam War. Consider this interview excerpt from a military veteran:

> My unit had originally been tasked with convoy duty, which was by late 2004, the insurgency was getting to be really effective at blowing up trucks and so all the convoys and civilian vehicles carrying supplies all over the country were guarded by [the] military and we were tasked to do that but then at the last minute it changed; I don't know for sure but I think that some of it had to do with the fact that we had women in the company. About late 2004 there were a number of stories in the news about female soldiers that had been killed in action and I think they might have changed us to a base side duty because of that. You know, this war was so well managed by the Pentagon, by the administration. I wasn't in Vietnam but I grew up with Vietnam, and it was every day on the six o'clock news – film

> from Vietnam, and it was often pretty horrific. [In this current war] the Pentagon didn't allow reporters free access. Everything was managed, [as a] matter of fact the commanding officer of my unit, [who] I got along with really well, he was a really good guy, [and] one of the few competent people that I served with … he mentioned there had been some reporters coming to the base. I stuck my hand up and said, 'I'll talk to them.' And of course, he knew my politics, and he said, "absolutely not." (AS, Interview)

AS believes that he had the parameters of his mission altered to avoid risk of bodily harm to women soldiers, and subsequent negative media attention. In the immediate post-9/11 era, "graphic photos of US soldiers dead and decapitated in Iraq … were refused by mainstream media, supplanted with footage that always took the aerial view, an aerial view whose perspective is established and maintained by state power" (Butler, 2004, p. 149). What does it mean for a government to shy away from politicizing or calling attention to the deaths of soldiers and the deaths or injuries of private military contractors, who often outnumber uniformed personnel? What does it mean when these acts and effects are not even publicized, let alone considered grand sacrifices worth folding into a narrative of national identity? This privatization, and/or denial of publicity, and the rejection of surplus that results, is centred on the body and the degree to which violence is embodied.

The idea of the body reflects a critical aspect of how sacrificial violence is applied, experienced, and how these constructions in turn inform the political. In Allen Feldman's (1991) landmark study of the Irish social landscape during 1969–1986, the body represents the central terrain by which historical and political memory, and action, are expressed and resisted. Through his analysis, the body's relation to sacrificial violence and the possibility of resistance to or reclaiming of sovereign power therein is revealed. Through the territorialization and classification of bodies alongside liberal spatialization techniques and apparatuses of regulation, Feldman argues that bodies are encoded and decoded within both public and private domains. Both state and non-state actors emit individualized and collectivized narratives of violence, informing the dialogically constructed identity of, for his purposes, the "Irish Self."

Feldman's (1991) focus is the Cold War state-building practices concerned with information warfare. This is reflected in practices of interrogation, detention, arrest, and house raids. Here, the state reduces itself to merely another paramilitary presence (p. 89), where the point is to collect information on individuals and communities rather than to ascertain truth, to charge and prosecute (p. 110). He shows how these categories are mediated by temporal and spatial properties: "crumbling boundaries between the inside and the outside, the private and the public" (p. 93). According to Feldman, "being done" is a right of political passage and indeed a sign of political maturity. As Feldman writes, male republican[4] biographies are inscribed considering this journey from an apolitical (feminized) subject to a masculine political subject: The body and arrest and interrogation experiences are the fabric of this passage (p. 98). Individual bodies became, in a sense, weaponized, indicative of historical collective memory and action. As a result, the "collectivized weapon endures; it is a hardening (reification) of the body. Thus, if embodiment is sublated by the weapon, if weapons become bodies, then bodies can be reciprocally metaphorized as weapons. Both 'hardmen' and 'stiffs' are bodies transformed into weapons" (p. 103). Prisoners literally begin to articulate their bodies as weapons and sites of political resistance and protest. Here, the weapon and the body become central and interchangeable political artefacts of paramilitary culture (p. 179).

For Feldman, and as the previous discussion on liberal and republican forms of citizenship attests to, part of becoming a political subject (citizen) requires these public experiences of practising, or at least implicitly condoning, violence among men. Feldman (1991) asserts that all prison revolts occur at the level of the body exemplified by "the H-Blocks [which had] their myths, local histories, performance spaces, and carnivals of violence, symbolic kinship, death rituals and animal totems" (p. 166). The Dirty Protest encapsulated techniques and discourses of the body concerned with the interior domains and excrements of the body used to counter their "colon-ization" and resist their sense of impending death (p. 181). In this respect, "the Blanketmen converted the interior of the body into a zone of trickery and a medium of communication with the world outside the prison" (p. 199). Here, the political capacities of the body are linked up with daily automatic processes of excretion. Unlike forced nudity, being forced "spread-eagle," medicalized mirror/cavity searches, and humiliation through infantalization, the Dirty Protest is representative of a *politics of resistance:* it is the one aspect of the prisoner's biology that the guards could not control, thus facilitating new political space to resist the sovereign demand of surveillance, rendering the prisoners outside the optics of the state's visibility. Excrement, bearing dirty historical relations and the trace of the other (p. 180), in addition to the basic biological make-up of the body, becomes communicative and expressive sites for both prisoners and guards. Through biological functions, the prisoners and guards are drawn into a relationship with each other.

They are intimately connected, even when not physically together: "the smell would stick to his uniform and to his body and was hard to get off ... the prison officers did feel defiled because it extended into their private lives" (p. 193). The prisoners and guards participate in the production of ideological imaginaries of the other – they do so vicariously and voyeuristically. Accordingly, sovereign violence is realized through the body, a product of shifting boundaries and political terrains of sacrifice.

Feldman's discussion of the body's relation to violence is illustrative for our purposes in the form of two analytical points: First, Cold War and post–Cold War interrogative tactics and rationalities were based on capture, arrest, torture, and the detention of bodies as a means to cultivate information, (and ultimately realize masculinized citizenship as well as the expression of sovereign power's expression of violence through their unequal albeit deeply intimate encounter or encounters with one another). Likewise, recall that Kahn and Agamben similarly describe the violent encounter/exchange as the space wherein sacred violence and state power converge and come into being vis-à-vis the body. Bodies matter in the performance of sovereignty. Second, these violent embodied encounters had a spatial-logical component that informed the realization of the encounter. This style of control or violence provided the space (prison/detention centre) and opportunities (through bodily encounters) to resist and/or alter the direction and pace of violence in their simultaneous claim of and resistance to sovereign power. Competing sovereignties – competing views on life and what entity has authority over it – are realized in the capture and enclosure of bodies. In acquiring information directly from the confessions of detainees, the materiality of resistance and agency is anchored firmly in the body, which exists within an institutionalized space.

However, US military policy, increasingly exemplified in the deployment of drones beginning in the post-9/11 era, bypasses the practices of enclosing bodies through arrest, capture, torture, and detention – refusing the confining spatial, embodied politics of the camp while redeploying its logic through air power – that are bound to arise in the act of encountering the ostensible enemy other. If "each of us is constituted politically in part by the virtue of the social vulnerability of our bodies ... exposed to others, at risk of violence by virtue of that exposure" (Butler, 2004, p. 20), then drones undo this risk of exposure, troubling the content of sacrifice in the context of a style of violence that normalizes it as a series of unending, distancing social routines. Here, technicians use drones not to confine and interrogate for purposes of acquiring truth, or mere information, but to erase the body; not to engage in a violent exchange but to bypass one; not to bring its victims into a *relation* of subordination but to simply eliminate upon contact. As Khan (2008) reminds us, "the United States literally does not know what to do with the captured terrorist" (p. 176). An interview with a lawyer during the height of the war on terror, who was working on drone cases for the non-profit organization, Reprieve, verifies this:

> We [Reprieve U.S.] expressly think that the reason drone strikes have increased is because Obama has essentially implemented a kill instead of capture policy. We are seeing a lot of that with ISIS [Islamic State of Iraq and Syria] and their conversations now being had about whether to detain alleged ISIS fighters and where to detain them and how to detain them and the government's been pretty clear that they are not interested in detaining them in U.S. custody, they prefer the countries on the ground detain them, because they got in so much trouble doing that beforehand, and in cases where that's not the case they're just taking people out. (SSB, Interview)[5]

Sovereign power then is denied in the absence of embodied and shared violence and is eclipsed in this minimization and ultimate erasure of the violent encounter. Girard (1979, p. 28) writes of a scene of violence, wherein two men come to blows, blood is spilt, both men are thus rendered impure. Drones offer a solution for containing the impurity of the other in that they offer the potential to eclipse the political constraints of confining and managing bodies. When asked about why the United States is so hesitant to detain people, the same interviewee responded:

> It presents a new opportunity to make the same mistakes that they did before [torture]. They would have to set out transparently what status they are giving these people, are they calling them enemy combatants, are they capturing them on a battlefield, who do they keep, where do they keep them, do they give them an attorney right away, do they have habeas corpus rights, do they have any constitutional rights, are any of the U.S. citizens, who are entitled to more rights than non-U.S. citizens … there's so much scrutiny, and they're under so much criticism internationally for Guantanamo, so now they're just killing everyone. It's much harder to tell who they're killing, and how many people they're killing because there aren't people on the ground, besides those who are actually being killed, to count the damage. (SSB, Interview)

Bodyless war means that there are no bodies to count the damage, nor are there bodies to manage captured bodies and no possibility of becoming impure through bloodshed or contact with bodies. There is no one to witness the carnage, or the consequences, or feel ethically attached to war's violence. This interviewee is describing a scenario in which the logistical and political *costs of bodies* – capturing, holding, interrogating, caring for, and processing bodies – are perceived to be too great, too costly, or too politically problematic. But sacrifice needs bodies: it must consume. Here, drone strikes circumvent these body politics, enabling the avoidance of sacrificial encounters and the collective turning away. The logic of sovereign power to erase bodies still exists, but this power does not reside in the body of the citizen-soldier, waiting to be made present, to be recognized, in the violent encounter. Instead, it has been reconfigured and redeployed through technological means on par with the logic that underpins governing at a distance. Curiously, sovereign power does not take comfort in a public display but is instead concealed in the moment of its revelation.

Summary

The outsourcing of war suggests some level of incompetence, inability, or *discomfort* with implementing war: States either cannot or do not want to engage in some aspects of war. Some might claim this is merely a logistical matter as war is now endlessly complex. However, I suggest that this discomfort or unwillingness signals an ethical breakdown between war's link to the citizen-soldier and thus society more broadly. In the social, political, and economic context of military privatization, distancing technologies such as drones are a natural outcome, as they extend the rationale of privatization and governing at a distance through technological means. In helping cultivate patience rather than wrath, privatization and drones work together to generate the dream – however implausible – of bloodless war, signalling the parallel breakdown of citizen-soldiering's relationship to sacrificial cults and idioms.

As the next chapter shows, drones are an extension of human senses, absent the physical, political, and economic risks of deploying and exposing bodies in battle. Drones are often relatively cheap, easily replicated, and sometimes disposable. This is in stark contrast to human soldiers, who are expensive to train, recruit, and retain and require rest, care, and innumerable other short- and long-term investments. Consider that humans are a literal burden to aircraft, which are designed to go as fast as possible but are in many ways slowed by the weight of a pilot. As bodies shift away from war, reducing risk for some but amplifying it for others, sacrifice, and its cultural, symbolic capital, is likewise displaced. This displacement, wrought by privatization, is apparent at two other related sites: in the rejection of the surplus contained in the aftermath of war's violence and in the desire for and practice of disembodiment. In the next chapter, I turn to drones to illustrate the abandonment of the citizen-soldier model of warfare. If we accept that the most valorized expression of sacrifice is contained within the image of the citizen-soldier, we must account for how the combination of the outsourcing and technologizing of war-making destabilizes and reconfigures this powerful process, amounting to a seemingly non-sacrificial or antisocial style of warfare. What work is the citizen-soldier no longer doing for the nation-state? In the imagining of the nation-state, the citizen-soldier represents the supreme expression of sacrifice, and therefore citizenship, which other citizens are compelled to aspire to. The figure of the citizen-soldier, making up the first group of people to receive regular benefits, signified ideally proper conduct for all civilians, as people who are seen as sacrificing for a particular body or association are regarded as "true" citizens (D. Burchell, 2002). However, the end of conscription, coupled with neoliberal flexible citizenship, troubled the content of military sacrifice and its ability to act as a check on violence. Rather than liberal democracy's promise of a decline in violence (we have fewer casualties), however, drone warfare – as the technological expression of the sociopolitical philosophy of privatization – is contributing to the expansion of more, increasingly invisible, and therefore meaningless violence. As I argue, drone violence gains credibility through a dialectic of visibility (the battlefield) and invisibility (the violence): By rendering that which was previously unseen visible – casting excess light on the battlefield to break up the fog of war – it claims to save and protect those lives that matter most: American and allied soldiers. However, the political and social landscape in which drone warfare operates and gains credibility, is dependent on the politics of trauma, the scene of sacred violence, and the identification of the witness all being made invisible.

What are the contested meanings of sacrifice – which I argue is defined by suffering, injury, and bloodshed creating surplus for the nation – in both the ideology and practical application of military privatization and drones? The paradox of sacrifice is clear in relation to the (potentially ever-expanding or unending) temporal dimensions of the war on terror and its subsequent series of related undeclared wars. Former calls for sacrifice, particularly those pertinent to the twentieth century, occurred in wars that provided more discernible, concrete timelines: with beginning, middle, and ending points, characterized by official declarations of war, and the signing of treaties, which signalled their end. Moreover, these were widely considered just and justifiable wars. Current wars, or a series of acts of "routine" military violence, have questionable justifications. They are nevertheless mobilized by a desire to protect life but lack clear parameters around the meaning of sacrificial acts. This is a *political* shift, which is amplified and concealed by neoliberal discourse geared towards generating greater forms of distance in war. In the next chapter, I question what it means for our understanding of military sacrifice when technologies generate a novel intimacy-in-distance and, thus, where witnessing, or the ability to have knowledge about and critique war, loses its publicizing potential.

Chapter 5

Sacrifice Lost

I believe I lost some of my humanity while I was in the drone program.

– ML (Interview)

NM: Humans are made in a certain way that when they start killing one another, it really destroys something in them that they can't restore. I think it's inevitable that that will happen. When we were in Ohio we were talking with a couple of guys in the Air Force, one of them had been doing this study of the drone pilots and there was some discussion in the Air Force about whether these pilots should be sequestered in some kind of base where the pilots wouldn't go home every day after work because essentially you've got people who are being asked to be executioners, every day they go to work and are asked to blow up somebody, and then they go home and play with their kids, "Daddy, what did you do today," "How was it," and this kind of stuff, I think that would make you crazy. And so I think the stress of this thing too has to do with flying these things at long distance, there's a few seconds delay in lead time, they can follow someone days on end and in way get to know that person, and then the order comes: "OK, kill them." That's different than if you're in a tree line and you blast … you don't really know that person, right? I think this is a whole other level of … it's kind of a worming into your own personality and eating it out from the inside if you're actually killing somebody you get to know …

BB: I guess only time will tell how the state will interpret these traumas … they can just say, "Forget you."

NM: I think that's generally what they want to do with all veterans. –NM (Interview)

Drones and the Future of War

What might drones tell us about the future of war? The controversy surrounding drones, at least for those who deploy them for military advantage, has become less apparent over the past nearly three decades.

This normalization has been largely thanks to industry efforts. Around a decade ago, there was a push from industry and defence and security experts to make drones acceptable in the mainstream. Part of how they achieved this was by emphasizing the life-affirming, biopolitical, and moral requirement to use them to protect US and allied soldiers' lives. Relatedly, the aesthetics of precision provided the basis for an ethical argument around preventing civilian casualties, and it did not hurt that this could also be pitched as a cost-effective means of warfare in line with widespread talk of austerity. Advocates claim that drones contribute to better and more humane war outcomes for all involved. Even as fifty civilians are murdered for every "terrorist" who is successfully assassinated by a drone (Kaplan, 2017, p. 163) the language of precision, moral superiority and humanity in war persists. Der Derian (2000, p. 772) captures this paradox of war's "humanity" through the notion of virtuous war. Key to virtuous war is "the technical ability and *ethical imperative* to threaten and, if necessary, actualize violence from a distance – *with no or minimal casualties* [emphasis added]" (p. 772). Virtuous war means that we are now interpellated into a drone imaginary, wherein, ethically, *we must comply.*

Criticism of drone warfare becomes a tough sell in the context of virtuous war, where US and allied life is the most important type of life and where drones help save said lives. This sense of helplessness – *how can we criticize drones when they are so life-affirming* – is part of a wider two-pronged problem, the first being that of witnessing in contemporary war and the second relating to the character of high-tech wars. Of the first, Richardson (2022) argues that traditional modes of media and human-rights-based witnessing have not yet accounted for remote warfare's distributed architecture of violence and the impact of material witnesses or non-human traces, agencies, and entities. Regarding the second point on the changing character of war, the idea that drones are the vehicle to satisfy casualty aversion compels a rewrite, according to Enemark (2023), of the dominant Clausewitzian view: regarding war as *interactive*, as "nothing but a duel on an extensive scale," "an exchange of violence" to a view of war where embodied dying and killing are no longer exchanged (Asad, 2007) but, more importantly, as I suggest, no longer *required.* This chapter focuses on how drones "remix" war (Gusterson, 2016) by altering war's spatial and temporal qualities (Kaplan, 2017), resulting in an intimacy-in-distance that undermines the archetype of the witness. In undoing the archetype, we privatize the politics of sacrifice, and we fray the threads of war critique, destabilizing the ability to know, remember, and prevent future war.

Despite promises of "never again" following the Second World War, and the attendant legal and political architecture to support the prevention of war, globalized conflicts of enormous devastation persist into the present while preparation for future conflicts is ongoing. Drones take centre stage as increasingly primary actants in many of these conflicts, yet as the introduction to this book conveyed, drones in and of themselves are not particularly new, radical, or interesting. As Caren Kaplan (2017, p. 161) writes, drone discourse removes most traces to the past, misdirecting historical, ethical, and political analysis. Drones are part of a longer reorganization of state violence. Still, despite technological advancements, their primary function remains relatively unchanged: to "observe, locate, and kill targeted individuals and groups" (p. 175). Drones, as the perfect tools of a casualty-averse society engaged in "everywhere wars" (Gregory, 2011) are an extension of the privatization of war, rendering violence largely invisible to Western publics with operations cloaked in secrecy (Richardson, 2022) even as the means, methods, and spaces of violence paradoxically expand geopolitically. That drones are frequently disconnected from their historical legacy in air power and air policing, and that they are part of a paradoxically shrinking and growing style of violence, eclipses something critical that liberal democracies have come to accept about war since the French Revolutionary wars and beyond, namely, that war inherently contains a political and publicizing essence (Enemark, 2023).

That this may no longer be compelling suggests a return to the past. What are the consequences of this for thinking about sacrifice and the future of war, given that, as previous chapters remind us, sacrifice is about humans and human-centred violence? It suggests at the very least, as per Richardson (2022), that we begin to account for non-human entities in sacrificial discourse, paying attention to human–machine assemblages and their reconfiguration. Mary Kaldor intuitively wrote in 1977 that "whereas formerly the weapon was the instrument of man, it now appears that man is the instrument of the weapon system" (p. 121). It also suggests that drones reflect a collapsing of functions; as they have become both the subjects and objects of sacrifice. The next section argues that drones signal the reconfiguration of the witness, or what it means to have knowledge about war and speak truth to it, through the politics of visibility and invisibility.

Intimacy-in-Distance and Undoing the Witness

Drone industry insiders speak endlessly about the limitless, favourable visual optics provided by drones – a visual terrain harnessed with the view to protecting soldiers while maximizing efficiency. What was previously unseen must be brought into view, and it is through drone vision that this is accomplished. Within this legitimizing discourse is a politics of visibility (intimacy) and invisibility (distance): While some things are being brought into greater view, catching the eye of "Western" publics and demanding attention, other things become invisible, going unnoticed or otherwise taking on a private quality. While remote and autonomous systems expose battlefields and bodies in time and space through endless opportunities and technologies for surveillance, such that we may see the entire world at once (Russell, 2020, p. 75), the *political* nature of the violence that the technique of exposure helps enact becomes increasingly invisible. In this section, I argue that drone warfare, because of its radical rupture of time and space, indeed signalling a rupture of *war-time* (Dudziak, 2013), relative to the endless cycles of violence that characterize postmodern war, renders the category of the witness, as a political archetype integral to knowledge and criticism of war's violence, increasingly troubled. In the undoing of the category of the witness, knowledge and criticism about war is also short-circuited alongside a reconfiguration of what kinds of suffering count as public and private. Those soldiers who labour to implement drones, on one hand, and those who experience the deadly and horrific effects of drones, on the other hand, are caught up in a shared logic (albeit with radically different material impacts): Neither can properly witness violence due to war's remixing (Gusterson, 2016). As such, cultivating shared knowledge or criticism of war becomes arduous. While remote and autonomous systems expose battlefields and bodies in time and space through endless mechanisms for surveillance, simultaneously the political nature of the violence that that exposure helps enact becomes increasingly invisible.

In thinking about drones and the meaning of witnessing, consider the Association for Unmanned Air Vehicles Systems International's annual convention, back in 2014. There, a distinguished military official, Kevin W. Mangum, indirectly commented on the issue of drones and the visibility/invisibility nexus when he joked that "nothing ruins a good war story like an eyewitness" (see Baggiarini, 2015, p. 129)." Here, the general could be interpreted to be saying that the enhanced sight wrought by drones *is distinct* from another kind of seeing: the act of witnessing. Put simply, an eyewitness, someone who has seen something and thus can speak truthfully about it, ruins a good war story (Baggiarini, 2015). Witnessing is deeply political as its focus is on ethical interpretations and the values ascribed to subsequent testimonial or truth claims. Accordingly, there is an objective seeing and a more problematic kind of seeing that is deeply political and so goes against the military ethos of keeping political opinions private and adhering to a strict hierarchy. Drone advocates claim, in so many words, that the technology is apolitical. What *they* – the drones – "see" is undisputably objective truth; there is only one single way to interpret the imagery or data, and the drone knows best. This universal knowledge wrought by drone vision is reminiscent of monarchical wisdom which comes via a direct path from God to the king. As this chapter shows, military planners and soldiers alike now look to drones to provide a kind of certainty and level of protection that we once hoped would come from God or religious convictions. Instead of holy wars inspired and directed and thus justifiable by God, drones guide soldiers into battle and give them the assurance of a prophetic vision in which political, ethical, or legal legitimacy might be lacking. This contrasts with what witnesses see, namely, a contested constellation of truths that, when deployed in the public realm, can be scrutinized and folded into the body politic and can ultimately call into question a single vision masquerading as a single truth.

According to Richardson (2022, p. 41) witnessing today increasingly centres on the visual, with eyewitnesses and survivors having unique cultural authority as embodied witnesses. Yet, Richardson claims, remote warfare, due to its asymmetric, distributed nature, and distance in space and time, requires a new assemblage approach to witnessing, which takes material witnesses seriously. For Richardson, material witnesses include spatially distributed non-human traces, agencies, and entities, or "machinic ecologies that archive their complex interactions with the world" (Schuppli, quoted in Richardson, 2022, p. 42). Agamben (2002) provides a temporally situated complement to Richardson's spatialized materialist perspective. Agamben defines the witness, on one hand, by the subjective potential for testimony and thus the production of truth as recognized by law and, on the other hand, as a survivor: "a person who has lived through something, who has experienced an event from *beginning to end* [emphasis added] and can therefore bear witness to it" (p. 17). What constitutes the beginning and end of drone violence is unclear, because their presence can be found in undeclared war zones, as a permanent feature of the sky whose dull humming can be heard by the people who dwell below; the timing of the potential strike is deliberately opaque but made to feel imminent. To witness an act of violence requires that the violence has a beginning, a middle, and an endpoint. On the other side, for drone operators, the blurring of the home front and the war front produces a shift in the spatial and temporal logics that we have come to rely on to understand the traumatic experiences of soldiers (Atkins & Seamone,2024; Dao, 2013; Fielding-Smith & Serle, 2014). In turn, this allows the possibility for a link to be made between trauma and war. Embodied trauma is integral to then initiating war commemoration and memorialization. Targeted communities too cannot refer to traditional war-time spatial and temporal logics because of drone vision's panoptic, and therefore persistent, presence.

"ML" is an outspoken critic of the US drone program, having worked on the distributed ground system:

> What people don't understand about drones, if you look at WWII: you had an airplane, the airplane flew, it came in, it dropped bombs and then it left. When it left, all these people could come out of the bomb shelters, and you know, the alarms went off, well in a developing country, you can make an effective bomb shelter out of sandbags, and so what would happen, is people would have the opportunity to run into these shelters, the civilians mostly, pretty much, and save themselves, and then the plane would leave and they could get on with their daily lives. (ML, Interview)[1]

In this excerpt, drones illustrate and enact techniques of government at a distance by loitering for unprecedented periods. Drones prevent "getting on with daily life" as there is no clear beginning or end to the violence. Theoretically always overhead, drones could strike at any given moment. Drones also successfully bring "populations into the terrain of state legibility and security so that they might become governable subjects" Adey et al., 2013, p. 3). Panoptic logic is the beating heart of the drone. Because of drone technology's permanent gaze, people exposed to ongoing surveillance and violence do not qualify as witnesses but, instead, are made governable subjects. Governing at a distance undermines the transformative potential of witnesses, both perpetrators and victims alike, whose experiences of drone wars have no clear temporal boundaries and therefore *cannot be*.

Targeted killing at a distance forecloses the embodied encounter between executioner and victim. As far as the technology offers more ease in effectively closing in on a battle space, it is nonetheless a space where the enemy is potentially intimately known. This is an intimacy-in-distance characterized by a need to "see without being seen," invoking a profoundly asymmetrical power relation. Yet, because "you (the drone operator) can see what's been done" its aftermath produces new kinds of trauma and suffering for those operators, unlike previous soldiers who "really weren't tied to it;" (TW, Interview). Pilots of manned aircraft simply dropped bombs and turned away from the damage.

This is not to say that traditional air force personnel did not suffer, but rather, their ability for intimate vision, and therefore to experience themselves or self-identify as agents of violence, was bracketed such that its impact might never be perceived unless deliberately recalled. Their suffering was limited by and contingent upon the technology at the time, with minimal visual imagery being imported into their post-war memories, these former pilots, coupled with a distinct view of air war, were afforded the chance to compartmentalize emotions and/or actions to whatever extent possible or desirable. They are afforded this in part because of the technology at their disposal. Piloting a helicopter, for example, over hostile enemy territory logistically does not permit an extended encounter.[2] At the same time, however, fighter pilots can be shot at. Their spatial and temporal experience of war is embodied and visceral. In this comparative vein, "LL" adds another key difference:

LL: The major difference is if you go to war and you go to combat, and I've gone to a combat zone, I've deployed overseas, so I've experienced both, when you go overseas, people have an understanding, so you know if you fly off the handle a little, people understand, "Oh you know she just got back." So you know, "give her some slack." But when you're talking about people who do this every single day, and they live right next door to you. … The public doesn't understand it. And even more so, the people in the military don't get it either. So you can work with people that are in the same uniform that are in the building right next door to you, and they don't get it either.

BB: So the airman's mission, it's all about physicality, and heroism, and sacrifice, so it doesn't apply in the same way to the drone folks?

LL: Nope.

While drone operators experience no immediate threat to their security, they must nonetheless confront their roles in ongoing surveillance and as perpetrators of possible lethal violence, even though the distance that drone technology affords not only minimizes their risk but also reconfigures their spatial awareness and the meaning of their presence in the battlefield. As Paul Rolfe, a former UK drone operator writes, drone pilots endure high operational stress: "You're absolutely there, you're in the fight … you're hearing the guys on the ground and you're hearing their stress, so when you finish your shift it's very odd to then step outside … it's the middle of the day and you're in Las Vegas" (Fielding-Smith & Serle, 2014). Rolfe describes a profound disconnect between his mind (which is on the ground) and body (which is absent). Indeed, drone pilots operate within different temporal boundaries: an unusual timeline of battle altogether, marked by their communications across different time (and war) zones, and an unsettling cognition of identity as pertains to space, which signals the emergence of a new normal for those who operate and engage in air power. As a visual practice, a form of "control without occupation" (Grayson, 2012), targeted killing is made possible through a fragmentation of time, space, and identity.

Consider the following remarks by Colonel Cluff in a *New York Times* article, where the authors posit that drone pilots are in effect perpetually deployed because their labour, while allowing them to engage the battlefield from the United States, means that they are constantly shifting from work to family activities. In contrast to the epic journey of Odysseus, who takes ten years to return home from war, Cluff claims, "having our folks make that mental shift every day, driving into the gate and thinking, 'All right, I've got my war face on, and I'm going to the fight,' and then driving out of the gate and stopping at Walmart to pick up a carton of milk or going to the soccer game on the way home-and the fact that you can't talk about most of what you do at home – all those stressors together are what is putting pressure on the family, putting pressure on the airman" (Drew & Philipps, 2015). Drone warfare, and the collapsing of the social space with the military work, indeed the collapsing of social identity that results – from soldier/warrior to/from father/husband – troubles the conditions of possibility for how one ought to primarily identify amid competing identities. The neoliberal demand for flexible forms of citizen-soldiering not only erodes the line between the beginning and end of a workday, as it does for many types of labour specific to neoliberal capitalism, but during and after work, it would also appear that one would feel out of place, disjointed, and altogether not quite right. Military violence is supposed to be exceptional – not just temporally but spatially too – but for drone operators that violence, and violent mentality, cannot be neatly contained upon clocking out. Violence becomes banal, even as the harms and mistakes continue to mount (Cloud & Zucchino, 2011). Furthermore, TW, a retired US Air Force (USAF) colonel and experimental test pilot, remarks,

> It's a different kind of stress, because flying a fighter is a lot of physical stress, and mental stress, I think the guys that are doing it from the ground, are probably even more psychologically stressed. I could see where guys flying on the ground, they have long work days and fairly high stress environments, they don't want to kill somebody that isn't supposed to be killed, how that could really be stressful, and especially a lot of the weapons today, *you can see what's been done.* (TW, Interview)

While remotely piloted aircraft pilots may stare at the same location for extended periods, manned aircraft depart the scene as soon as possible. Furthermore, drone operators are not afforded the ability (whether desirable or not) of mental compartmentalization. "VH" has experience as both an American soldier and a private military contractor.

> VH: The biggest thing after 9/11, depending on who you were and what job you had, you could call home every day. Or if you wanted to, depending on what you were doing, what your job was, but you could call home, or have email. You had email available, even if it was 20 minutes a day. Vietnam, you waited for weeks for a letter, or WWII, even desert storm was primitive in that regard. … I think the rise of the drones is horrible; it freaks me out. What just makes me sick is I don't know how people can, somebody sitting in Arizona flying a mission to blow somebody up in Yemen, and then going home at night. There's Air Force pilots that fly out of South Dakota, they fly, then come home.
>
> BB: Why does that freak you out?
>
> VH: I think that being in the environment – I was in Iraq and Afghanistan – when you're over there, it's total immersion.[3] You don't have a choice, and everybody around you is all involved in the same thing, more or less. There are no distractions. I think it's a lot harder for the families at home than those of us over there. For us it's cut and dry, you don't have to worry about taking the kids to the baseball game. Mixing the two just really freaks me out. It's bad enough as it is, if you're sitting in Afghanistan. There are people back here, at Fort Huachuca, that have PTSD [post-traumatic stress disorder], and they're here in the States. Sitting in the States. *So, from a military cause of PTSD, you really shouldn't get that if you're not in a war zone.*" (VH, Interview)

Violence streaming via live video feed generates an intimate bond between soldiers and targets, creating an intimacy-in-distance that is defined by war's remixing of spatial and temporal boundaries. Because drone warfare is legally upheld by the continued reauthorization of the 2001 Authorization for Use of Military Force, which allows the United States to deploy armed forces against those who are thought to have contributed to 9/11, or so-called associated forces, the war on terror is by definition an unending war that ruptures the logic of time and space that characterized previous modes of violence. To witness violence, to be regarded as a subject with the potential for truth-telling, is to experience its closure. But drones and the war on terror more broadly, as part of visual practices (Grayson, 2012), offer limitless and expanding forms of surveillance and targeted killings. This leaves perpetrators and victims caught in the same logic, albeit with different material implications. As drone war's panoptic logic moves violence from exceptional to banal, neither subject can witness. Neither can "know" war's violence; thus, neither can leverage this knowledge to force a critique of war. How can war be remembered in this context? Is it preferable to forget? I will return to this in the conclusion. For now, let us trace how practices of in-visibility inform drone warfare.

Make In-Visible

The main advantage of drones is that they are thought to protect the warfighter and ultimately save soldiers' lives, but this protection is only possible to the degree to which drones can make visible that which was previously unseen by dislocating and extending the scope of the military gaze (see Baggiarini, 2024). To make live is to make visible. However, it is the invisibility and the anonymity of drone violence that deserve equal, if not greater attention. Enhanced sight and interpretation of data wrought by drones are distinct from the politicized act of witnessing: an act of critique through the authoring of a counternarrative of a violent event. However, military practices are now deeply privatized and rendered invisible, sovereign violence is no longer the legacy of the citizen-soldier or nation-state such that, in a way, we might imagine how drone violence emerges as a "non-event." State terrorism benefits from the privatization and depoliticization of the witnessing of the event through a minimization of those who have access to it, but furthermore, through the language of technology and security; drones, as both the subjects and objects of sacrifice, re-classify the witnessing of the event. The non-event of state terror produces terror without witness and without premonition, invoking the omnipresent power of God – blending divine retribution with profane catastrophe (Rupka & Baggiarini, 2018).

The acquisition of uninhibited vision in the form of a legal and social erection of a swift and potentially permanent surveillance architecture is thought to be the final stone in the construction of a completed edifice (Rupka & Baggiarini, 2018) – the eventual complete protection of the American soldier. "The drone as a concept and as a system, as an actual system, is really a surveillance system, that's ninety-nine per cent of what it does, it's used for watching and targeting" (PC, Interview). Drones, as enabling architecture of panoptic surveillance, give the impression that you can lift the fog of war, you can change this century-old idea that the opponent might surprise[4] you. "He cannot surprise you when you have him under constant surveillance" (NS,

Interview). In this era of virtuous warfare, life is the primary concern. Drone vision protects life by making visible to eliminate the element of the invisible, or surprise:

> I was in Iraq in 2004 and they had them [drones] around the base all the time. They were using them for surveillance around the base. And of course I saw them where they launched them from and that kind of thing, all the ones I have personally seen are all small. They had the little ones with cameras and I'm sure they flew them outside the base as well. From a technology perspective, they're great. Nobody's at risk, you can get more information than you can when you have people out there driving around, or walking around, even in an aircraft ... the system that I worked on came into play directly as a result of an attack on a base in Afghanistan, where they just got ... nobody saw them coming. It was a small little base, and a lot of people got killed. And they decided they needed something to give early warning, so there were camera operators that are 24/7, looking, operating these camera systems, but that's different. Those things were used for targeting also; I mean occasionally, I saw it a few times. They found somebody putting in an IED [improvised explosive device] in the middle of the night and then called in, you know, an Apache helicopter or something to go blow them up. (VH, Interview)

Accordingly, the IED represents the tool of the terrorist, which drone vision helps eradicate. The historical context for this aspiration for an unending military gaze is a post-Vietnam identity crisis around how to avoid casualties, defeat, and subsequent unwanted publicity when things go awry. This crisis provided the technological incentives to prevent further loss of life in future missions, making the protection of life the centre piece of military strategy and its protection of life then the measure of a mission's success. Indeed, as "PC" confirms, these drones are operating in places where there are no soldiers. Moreover, "there are no troops [in Yemen]. We are not at war with Yemen" (VH, Interview). Fragmented, decentred, and disembodied military power, not its reduction or minimization in terms of scope, is the central component of postmodern violence. This goal is accomplished, I argue, through drone vision: establishing the potential for enhanced and unending sight to protect life.

The synthesis between technology and protecting life was made especially compelling in light of the 1991 Gulf War, which was widely hailed as a highly efficient success in part because the American casualty rate was unprecedentedly low. This was attributed to a historically unparalleled application of military technology (Shimko, 2010). Consider this interview statement from a veteran who served in the Cold War, Iraq, and Afghanistan: "Kosovo is I think the best example of a technology war. People sat at home and lobbed missiles at Sarajevo.[5] And people didn't get their hands dirty as much. The Gulf War, earlier, was I think the only reason it was different, was because it was so easy in the end" (VH, Interview). In this instance, high technology first developed and applied in the Gulf War allowed for people during the Kosovo War to "sit at home" and still achieve victory while keeping their hands clean. "Sitting" challenges traditional imaginings of citizen-soldiering, namely, that minimal physical fitness, effort, or endurance is required, while the use of "home" portrays war as risk-free and comfortable. This new model of fighting, namely, where soldiers engage in war yet avoid the battlefield, has a history that is intimately tied to America's defeat in the Vietnam War[6] and related emasculation after a performance that was regarded as rife with disciplinary problems and an abysmal demonstration of military ineptitude (Masters, 2008).

During this time, while dealing with the political and symbolic consequences of defeat, American military strategists also had to address the disciplinary problems associated with the transition to an all-volunteer force, a system that does not always attract the most skilled or technologically perceptive personnel.[7] As one drone whistleblower explained, drones are now the preferred military tools because "it's politically expedient. I also think that some of it [is to act as] counter measures from all the protesting that was done during the Vietnam conflict. People would be up in arms, and complain, if they knew. We still glorify the indiscriminate killing of others; it's almost traditional, sadly" (LL, Interview). Consider also this following exchange, where it appears the drone technology itself, combined with the ideology of casualty aversion, is accredited tremendous power, such that it may shift the nature of military missions- how they are conceptualized, and prioritized- in a dramatic way:

TW: I think all the press that's been out there has given a negative connotation to the use of unmanned aircraft through weaponized stuff; I think the one thing that people miss is that these systems are *a lot more precise now* than manned aircraft used to be and in fact in some instances manned aircraft are right now, as the weapons get better and even more precise, you're going to end up with a lot less collateral damage than we've had in the past.

BB: ... It does seem to change the nature of what kinds of mission are considered acceptable and appropriate.

TW: You're perfectly right, you would say you need either a very good way to legitimize causalities and that would raise the bar, or you would go for missions that guarantee very low casualties, air strikes rather than boots on the ground, so you have impacts on the selection of missions, on the strategies of the mission, impacts on tactical level, so this casualty issue is very important on a whole range of issues for political as well as military people who have to fulfill the missions. (TW, Interview)

"SG," a drone lobbyist and member of the Association for Unmanned Vehicle Systems International, further confirms this (contentious) shift in how to define the enemy and therefore the mission parameters – note the primacy of economic language ("cost–benefit analysis") used to rationalize his perspective on remotely piloted aircraft (RPAs):

From what I've heard recently is that there are more new pilots being trained for RPAs than for the real ones. And the question is a cost–benefit analysis. How does it help to fulfill the mission? The mission is changing. Today we're undergoing major changes what that mission is. Who is the enemy? What is their rule of engagement and how do we defeat it?" (SG, Interview)

Technological innovation provides the rationale to make some missions more worthwhile than others, those thought to be "too risky" because they may put troops in harm's way versus those that pose little to no risk. This profoundly limits what kinds of military solutions might be afforded to state and non-state actors looking to negotiate and problem-solve in conflict zones, reshaping the very possibility, and meaning of human-centred interventions. Technology also provides the solution to foster this distance between "us" and "them," thereby protecting (our) lives, and theoretically minimizing risks to the unknown others.

In the following, in the first excerpt, the desire to protect soldier's lives is directly linked to the haunting of the defeat in Vietnam – a "history of lost wars." In the second excerpt, from a different interviewee, a retired USAF colonel and experimental test pilot, the saving of lives, including Americans, as well as in terms of preventing collateral damage of unknown others, is attributed to the ostensible precision of drone technology. Finally, in the third excerpt, a Defense Advanced Research Projects Agency (DARPA) employee ("ST"), when asked to define a drone and suggest their benefits, tells me that drones are unmanned vehicles that allow for greater protection of soldiers while extending their reach via enhanced visibility. "NS," an academic, explains:

The basic problem is that Israel and the US have this history of lost wars, more recently where they had casualties, and the casualty factor had been way more important in the US and Israel than in Europe, because we [Europeans] did not have comparable scenarios like Vietnam etc. *The marketing in the US is based on: drones will save soldiers' lives.* And that's the major argument, it's the most important argument in Western democracies, especially when you have this history of lost wars where you had high casualty rates, so this way of marketing, I saw it very early in the US early in 2004 in Washington where you had billboards that said, *"Those UAVs [unmanned aerial vehicles] will save soldiers' lives."* (Interview[8])

More than half of the pilots are unmanned pilots; the Air Force has really embraced that particular technology. Look at how many predators and global hawks are out there and what they're doing, and even on the small side, *all the services are embracing the smaller UASs* [unmanned aerial systems] *because they're so capable of saving American lives,* I know the army, it's become a part of their whole, concept of warfare. ... If you compare it to other wars, the collateral damage or collateral casualties has just been less significant. And if you look at where we are today, throughout the whole spectrum, that is probably the biggest concern of everybody, collateral damage, so that is why everyone is working really hard to make weapons more and more precise, so that you don't have collateral damage, but I think history, if you look back in history, I think you'll find that in today's world collateral damage is significantly less than it was even ten years ago, five years ago, I believe that's a concerted effort on the part of the military and everybody else to make that happen, because no one wants to think that they dropped a bomb and killed, you know, other people that they shouldn't have, but that's sort of the price of war, if you get into the nitty-gritty philosophy of war. (TW, Interview)

A drone is a popular term for an unmanned aerial vehicle (UAV), and *its primary advantage is in keeping our soldiers out of harm's way and extending a soldier's reach.* For instance, UAVs equipped with cameras can provide information about an area before a soldier must enter it. (ST, Interview)

In today's globalized wars, American military planners measure success as a function of their ability to minimize the introduction of ground troops, as an effect of outright casualty avoidance (Mandel, 2004, p. 1) thereby establishing a positive correlation between what defines the parameters of a mission, and what counts as political victory as seen in the protection (wrought by technology and mastery of it) of Western and/or North American life. As one interviewee asserts earlier, "[Collateral casualties] is the biggest concern for everybody." Drones are gentler and kinder weapons systems. Nevertheless, the question remains: What precisely is a drone? The answer is not entirely obvious, as made clear through interviews from both critics and advocates of drones (advocates/lobbyists will typically reject the term *drone*). However, what is clear is that a drone is simply one piece of a widening architecture of panoptic violence, which is anchored in practices of surveillance and a philosophy of targeted killing, where decisions are made based on "signatures" of military activity and/or mundane calculations:

> Now as I look at it, what's a drone? A drone is an unmanned aircraft. Now that drone is out there to do either surveillance, or it could be armed with bombs and missiles and so forth. Now the near fact that they removed the human being from that air vehicle, creates a different set of psychological issues, and *one of the things the Western world is always respected is human life*, so if a drone is shot down, there is no human life sacrificed; killed or captured (prisoner of war). So that's one thing that people who are being [watched] that thing has been taken away. The second thing is long duration; you can have a drone hovering for 24 hours a day, which is not practical. But my view is that if you look at the warfare over even before the First World War and go beyond thousands of years prior to that, human beings have always found a way to have a technological edge over the enemy. Drones are just a step towards that evolution. (SG, Interview)
>
> I think a lot of people in the military don't actually know what a drone is. They know their specific role, they imagine a camera maybe flying a plane, but the entirety of the system, the surveillance system, for the military, in reality a lot of people in the military especially the young kids who were hired as imagery analysts don't actually know, it's not in their job description. (PC, Interview)
>
> "IG," a robotic aviation expert, defines a drone as:
>
> *robots that fly; Two dimensions, it's a flying robot.* It knows how to fly, you give it high-level commands to go fly where you want it to fly. No one pays you for that, apparently. So it's all about the data. Now you've got access to a third dimension, what do you want to know: now you can go collect in the air, so what do you want to know? The military usually wants to know, is looking for bad people doing bad things, or pattern of life, those all tend to be things in real time, look over here, look over there kind of stuff. What you want is a robot that flies, that is safe, and that collects data for you. (IG, Interview)
>
> What I focus on is military drones, actually are often called, an unmanned aerial system by industry because they like to sell the idea of these robots, and what the military calls remotely piloted aircraft, and I actually like the military's title the best because in fact *the military drone is not in fact a robot at all*, it is a combination of a couple hundred people who are networked together from around the world, who use a vehicle that

> happens to be unpiloted, they use it for surveillance, for tracking people, and occasionally for killing them … the drone itself is just an aircraft, within it there's a lot of sensors: FMV (full-motion video sensors) there are electro-optical sensors, there are synthetic aperture radar sensors, aerial precision geo location sensors, there's transmission equipment, and each of them requires a manager, so you have multiple imagery analysts, multiple full motion video analysts, and of course the pilot and the sensor operator, you have radio technicians, so once you add all these up, each of these pods that have three to four aircraft require hundreds of people to keep them operating, and that to me is the drone – is those people – and what they're involved in is not killing but surveillance, that's their primary task. (PC, Interview)
>
> If you really understand the system, like a predator, it's a pretty simple aerodynamic airplane, it's not real complex. (TW, Interview)
>
> It's nothing more than a model helicopter, or airplane that have been around for ages. So many people have had experience flying model airplanes when they were kids. That's what it is right now; It's nothing different. (BD, Interview)
>
> Here's the thing, like everybody talks about it like it's just a plane. It's just this unmanned vehicle that flies somewhere, but what they don't see, is that some of the decision-making processes are algorithms, is that the future of where we're going? Are we going to look at trusting an algorithmic definition of what an enemy combatant is? A bunch of data is put in, a calculation is done, and I'm sure it is overseen by human beings somewhere in the links and the chains, but not overseen, overseen is not the right word, I'm sure that it's perhaps viewed by human beings before the strike takes place, but really we're basing it on an algorithm. (LL, interview)

These divergent and contradictory definitions differentially understand the simplicity or complexity of the technology itself and the place of bodies[9] in operating it, signalling a disagreement about not only the role of humans but also how simple or complex, autonomous, and/or animated, the technology is and ought to be. It is perhaps not surprising that the interviewees who advocate for drones (IG, TW, and BD) depoliticize drones, referencing them as if they exist in a social vacuum, while those who are more critical (PC and LL) emphasize themes of embodiment, politics, and drone warfare in practice. Regardless of the domain, the technological system that is a drone joins two distinct processes in one machine: "accurate" weapons systems, and speedy data transmission. As IG put it earlier, "it's all about the data."

A drone, irrespective of its size or purpose, is made up of fundamentally basic technological elements, although the technology is dynamic and, as such, in some sense, alive:

> One of the things that I really like about drones, is that *each part evolves*: Let me unpack that a bit, so when I talk about what we do, we provide an "airborne data acquisition architecture." We're that airborne acquisition piece of the industrial internet. Anyone can go to Best Buy today and pick up a drone, whatever you pay, different capabilities: what's significant is part of the safety thing is telling each other where we are and there's transponder technology that is coming that's now coming into both manned and unmanned systems – that airspace access piece is the next thing – globally because we have all these devices that anyone can put in the air so it now behooves us to put transponders on them so that we can avoid them. The big thing is knowing that they're there. (IG, Interview)

Drones synthesize (a) advancements in flying and navigation, such as "sense and avoid" or "beyond line of sight" (BLS) technology; (b) autonomy (to what extent the human is imagined to be involved in its deployment is increasingly ambivalent; future drones will communicate with each other but not necessarily confer with humans); and (c) surveillance technology, such as cameras. "UAVs' can be configured with various types and sites: optical, retro optical, thermal, so you can see different spectral ranges" (GB, Interview). Therefore, the fundamental power of the drone can be reduced to its capacity for enhanced, prolonged, and uninhibited vision; it is, effectively, an eye in the sky thought to "lift the fog of war." The RQ-4 Global Hawk, a long-endurance high-altitude surveillance and reconnaissance drone, is an example made popular by mainstream media. Drones, as a mechanism for unified and uncontested representations of truth, have, according to their proponents, the built-in potential to ultimately overcome human error – the human's inability to see clearly (objectively) without being blinded by politics, or blatant incompetence: President Obama echoed the importance of permanent vision in an address to the nation on 14 December 2015, which was directed to ISIS leaders: He simply said to them: "ISIL leaders cannot hide" (The White House, 2015).

This reflects an "aesthetic of transparency" (Hall, 2007) where, since 9/11, drones have been used in 95 per cent of the United States's targeted killings, thus normalizing the surveillance practices that unfold from it, because a targeted killing, in theory, requires a history of drone footage and corresponding signals intelligence before it can occur. Much of the data captured by the drone result in mundane images. Yet drones do not just simply capture things in real time; they also create a backlog of events, flooding historical data into the present, to be interpreted. This surplus of (mostly) banal imagery, however, is the basis of intelligence that may authorize a strike. That the everyday life of the drone's gaze can instantly produce both dull and disastrous results shows how state and non-state terror operates around different logics of legitimacy, to be sure, as well as around notions of visibility, witnessing, and legibility.

While rarely deployed under President Bush, President Barak Obama dramatically increased the use of drones. The air force, despite severe problems with burnout and retention difficulties,[10] now trains more drone pilots than fighter and bomber pilots. Drones are seen as the logical culmination of weapons development, what is often termed a natural progression in aviation that has revolutionized the conduct of war. Yet the services have embraced drone technology due to a desire to protect life by making visible what was previously unseen:

> [Drones have] an impact for the better; the last thing [our company] did was Afghanistan; that was a type of warfare. Now we're deploying troops to Latvia, to stand at the border. And I've done that job! I did that as a young guy in Germany. It's the same thing. *So that's a different warfare: peace.* The Minister of National Defense is preparing to send troops into Africa for peace support to the UN, that's a different problem. But they have one thing in common: that is, *the better they understand the environment, the more effective the operation is, and the safer the operation is.* There's now this ubiquitous, cost-effective capability that you can deploy in any of your operations, it does change decision-making, and because it's ubiquitous, it's not just Waterloo, right? General on the Hill, right? It's not just the General looking down, things are more distributed, people can get information at a lower level, as the mobile internet has changed our world, the mobile internet with a third-dimension view, changes the operation view. So, you're a patrol, just getting food aid, and there are bridges and, you know, crappy roads and the occasional bandits, and other stuff but now you have this *view that changes the world.* I think there's a positive aspect to this that you can't understate. (IG, Interview)

Drones are ubiquitous, a seemingly unstoppable and universal solution to various kinds of complex but ill-defined social problems (war and peace are not opposed but, instead, overlap as similar "problems"). As such drones are the perfect tools regardless of the social specificities. "A view that changes the world" is possible only with nearly perfect intelligence and, therefore, (ideally cost-effective) ownership of one's environment. Regardless of the "problem," whether it is peace or war, a view that is radical enough to "change the world" of the individual war fighter must offer more than a single, homogeneous, and, therefore, isolated vantage point; it must be one that not only looks down from the hill but also above, around, beneath, and through. To complicate matters further, this view's heterogeneity must somehow be synchronized within the non-hierarchical network. Moreover, one must be able to acquire that vision without risk, or "sticking their neck out." A UAV consultant explains:

> I did work on a bid for a manned portable one [drone], folds up and goes on your back, if you need it you throw it up, it has a small TV screen, and fly it over the hill and you look at what's there, the company commander can go and look at it and go "Ah ok, cool," and it flies back and falls on the ground. It's made out of carbon fiber so you roll the wings up and put it back in its tube, where it's recharged by a little solar shell. So, it gives you a chance to look over the hill without sticking your head out. You can drive it around and look at whatever you want. And its manned portable, and at a very low level, so it's deployed down to the soldier level – the tactical level – and it gives them that chance, we're coming up the town, what the hell's in there, send this over, it's quiet, it's black, it's designed to look like a bird, it's much more organic looking. (GB, Interview)

Gone are the solitary generals and their singular visions. Visibility and invisibility, exposure, and erasure are mutually enforcing ideas explained in part by the desire for enhanced optical ranges and thus the presumed life-saving potential wrought by drones. The theme of acquiring uninhibited vision for purposes of lifting the fog of war and avoiding being surprised by the enemy is marked by a further problem for military planners to overcome, namely, of how to enhance and accelerate the speed of data transmission such that the clarity of communication, and focus of imagery, can be effectively utilized to enable the dominance and precision of the disparate yet ultimately unified American military gaze(s). Superior vision for situational dominance is opposed to a simple top-down view from the hill. It is temporally bounded insofar as it is constant – occurring in "real time"; it is deliberately fragmented to enable many simultaneous gazes and perpetuates a multidimensional theory of superior vision. It is positively correlated in security experts' discourse with saving lives. Life itself is understood through this new kind of vision as that which is the drone's beneficiary. We owe life to the drone.

Specifically, advocates of drones and industry insiders attest to how the now ubiquitous presence of drones is underpinned by the perceived potential to permanently alter the war theatre such that the concern over human casualties might be negated. Within the drone industry, however, there is agreement that whatever the drawbacks might be, drones are integral to the outright saving of life, if not, at a minimum, the enhancement and protection of life. Drones allow humans the opportunity to avoid some of the "dangerous, dirty, and dull" tasks, and because of this, they are argued to be a force for good. Consider this testimony of a chief executive officer of a robotic aviation company, whose company is described as a leader in delivering airborne sensing solutions. He describes the rollout of drone technology in Afghanistan:

> A significant emotional event was in 2008, I was standing by the side of the road at four o'clock in the morning. Standing outside the tent in Afghanistan, we were going up to the site to become operationalized, and we were doing a convoy escort. And the convoy escort, just at that moment while I was waiting for the other guys, it drove by, and I looked up at the convoy commander, and it was a significant emotional event because, at that point we were losing an incredible amount of folks to IEDs, and now, all day, this guy had this unseen angel flying above him, looking ahead, and somebody else had a view into what he was doing, and so we were able to focus the energy of the organization into making that successful and safe. (GI, Interview)

Drones act as innocent and caring guardian angels, symbolically perched on the shoulders of soldiers working for the greater good so that these soldiers might ultimately live. Unlike the fallen heroes who came before them, who ultimately perished because of poor vision and an inability to control the element of surprise (which presents in the form of improvised explosive devices), soldiers of today are guided into battle not by an all-knowing, singular being (God) but by a fleet of all-knowing and all-seeing, clairvoyant techno-demigods. In this way, we might identify drone technology as constituting the ground for the "re-sacralization" of warfare. Drones have sovereign power, a sovereign presence, guiding the civilized into battle. Through its life-saving ability and enhanced vision from above, drones confirm that God is on "our side."

According to the first interview subject, drones effectively minimize the confusion wrought by differing interpretations of an event that are bound to occur when imperfect humans are left alone to their own devices. As one interviewee, a veteran of the Canadian forces and a current UAV consultant described, UAVs provide an uncontested visualized form of objective truth, without the messy burden of interpretation. In this view, drones

> can do some reconnaissance that we would normally send people out to do, it's very efficient at that, UAVs' can be configured with various types and sites: optical, retro optical, thermal, so you can see different spectral ranges, they're very hard to detect, they are very small, they don't make a lot of noise, and they're not reflective in any way, they're very hard to see, they provide real time … and you can record it, which is a key part, because instead of someone coming back and saying, "I saw x, y, and z," you can say, "Well let's look at the date." The commander can say, "show me that hill we're supposed to attack, I want to see the UAV footage," and he can look at it himself. From the point of view, it doesn't endanger anybody; it provides better intelligence that can be independently reviewed by other people, versus one person says, "I went and saw it, this what I saw." OK, it depends on their description skills, and what they did see, other people may have described it differently, two different guys might say two different attitudes, two different responses, *whereas the UAV can review the film.* It can see just as well at night as it does during the day, which people can't do, and it's in spectrums that people can't see. (GB, Interview)

For this interviewee, the assumption is that while more people will be able to see the footage in question, there will be implicit agreement in determining the meaning of the footage – the qualities of the desert landscape or urban setting being filmed and the intentions, actions, or future potential actions of any people who inhabit its surface, their thoughts, and their interactions with others, should now be self-evident *because the drone will make it so – the UAV can review the film*. While the drones – here attributed with a kind of personhood, capable in and of themselves of reviewing film – are more efficient than people at "seeing," people nevertheless see *through* the drone as far as they are meant to analyse the footage after it has been captured. The drone's eye is an extension of the human's eye, which is here thought to be weak and fallible. This gaze is nonetheless made stronger when inserted into the drone's technological apparatus. Through the pausing, zooming in or out, and/or slowing down of time, the eye surpasses all other sensory body parts as that which can determine the outcome of a mission.

A more critical interviewee contradicts this and suggests that honest interpretation of footage is fraught, at best, and impossible, at worst:

> If you're just watching let's say a temple or a mosque or a church, or some such thing, and you can just put it above fairly easily just like you would

a security camera, pretty easy. When you're trying to track people on the ground that's when it becomes much more complicated, more problematic, not so much because of the drone, but because of the technologies it uses to locate those people. It doesn't actually work. If you set up a camera, and point it at something, especially a low-resolution camera, which is actually what they use because they can't transmit it at high speeds, it's actually easy for me to fool that camera if I know what I'm doing. On the other hand, to fool a human being who's watching, is actually much harder. And true you can record the whole thing and play it back and back again, imagery can be very subjective. I mean there are things it does better because it doesn't fall asleep. *But given the low quality of the resolution, it's not objective, actually it's not so much that it's not objective, it's not very accurate, and you have to do a lot of interpretation.* (PC, Interview)

Moreover, when asked about misrepresentations of UAV technology, a UAV industry insider confirms that the technology itself is flawed in that it does not, and cannot, live up to its perceived potential as claimed by marketing campaigns. He, in line with the preceding interview excerpt, believes that perception carries tremendous weight. (However, it is worth noting that this response was initially a reaction to the belief held by some critics that drones could be used for spying on neighbours and/or invading personal privacy in general. His claim – that the technology that is commonly in circulation among citizens is not good enough to accomplish this kind of activity – is deliberately taken out of context by the author to, in some sense, challenge the content of the interview to show how the argument could similarly be applied to claim that the technology is not very good *full stop*).

I think there is a lot of exaggerated [claims] laid out by the manufactures, and the capabilities of their particular aircraft, I've seen some things that I flat-out know that they are saying they can do, and they can't … why do you think every camera that you see, the movie guys, they all have a tripod? The aircraft is bouncing around so much, when you zoom in you can demonstrate with your iPad or whatever camera you want- just sit down there and try and hold it steady on a zoomed in. … For instance, you're trying to read a bar code with your hand-held camera, see how hard that is to do. The only way you can really do it is take a snap shot, there's no video. That's the same thing when you, Aeryon has a new, with a new 30- or 32-to-1 zoom lens on it, and it's difficult, because especially if there's any wind at all, if you have a rotor, with one of the propellers slightly out of balance, it's gonna set up a vibration, so perception, I've fought this for 30 years, this perception, what they can see. Yeah, I can see you, a distance

> away, but I can't look in your eyes, see your emotions, you're moving, the camera's moving. The other thing that always gets me is that everything has to be rendered in a video, you're looking at so much, it's gotta be processed, it's perception … you can show someone what the capabilities really are, and it's not what they're imagining. (BD, Interview)

The misunderstanding about the nature of drone vision is revealed again in a third respondent, a former signals intelligence analyst. He claims:

> My task was to find high-value targets and support the mission by doing so through a series of signals intelligence, I would work directly with drone technology, the more official terminology is used for it is RPA or UAV. The basic idea was that I would help find targets and *the drone would go looking for them.*

Who, or what, precisely is going and/or doing the looking, and wherein agency lies is unclear. The original point of this gaze – if it is to be located at all – is not immediately self-evident. Again, the dynamic between the operator and the drone reveals an understanding of vision as being a supplementation of sorts; the drone, here attributed with its own agency, aids the person in seeing what they cannot see on their own. Finally, another interviewee, a retired USAF colonel, drone lobbyist, and experimental test pilot, attributes the precision of today's weapons systems due to better synchronization between weapons and data transmission, alongside enhanced GPS, and vision capabilities:

> Weapons have gotten a lot more sophisticated, and the night capability has increased significantly and weapons' precision is just phenomenal in today's world, guys, both for manned and unmanned aircraft, the precision bombing is just phenomenal, when I was flying F4s and dropping bombs in hard sight I was lucky to hit the earth; in today's world, you can drop bombs from pretty high altitude and still get right on the mark.
>
> A lot of the weapons today, *you can see what's been done*, you know whereas in past wars you haven't had the information, and the internet, and digital cameras and all that kind of stuff that you got today, *the guys really weren't tied to it*, they were either seeing it first hand or people weren't seeing it, and now, everybody sees it. Same thing, it's the same bad stuff – but the proliferation of information, that's I think part of why people are, now they're seeing the horrors of war, before they just sort of read about it.
>
> There were like 20,000 plus ravens, used by the army, there were over 20,000 of those built, and they were used by a lot of people in Iraq and

> Afghanistan, those people, soldiers, and marines, and air force guys, come back, and a lot of them get into the public safety applications, and they go, "why can't I have something like this" as a law enforcement entity, where I can look behind a building, whereas in Iraq and Afghanistan they're looking behind a building to see if there's bad guys there with guns, or if there's not bad guys there with guns, or, "Why can't I have that here so I can see if there's a bad guy behind that building with a gun?" (TW, Interview)

In the last line ("why can't I have that here") drone vision is thought to be an entitlement: if the possibility exists to see "if there is a bad guy behind a building with a gun," it is a moral failing on the part of the state to not provide that soldier (or police officer) with whatever tools are needed to enable that, particularly given the mandate of ethical war. Doing so would be compliant with the dictates of virtuous war both at home and abroad. However, as this interviewee gestures to earlier on in the excerpt, there are drawbacks to this vision. Whereas "in the past" pilots could not be tied to the act of bombing in any kind of intimate way, there could be no immediate post-bombing analysis, no post-bombing re-inscription into the event, and no plethora of media and information technology to catalogue events into a narrative of "what happened" thrusting the past into the present with a click of a button. In drone operations, pilots still are not tied to the act of bombing in that they simply navigate the plane and ensure it is operating safely and on target. But their counterparts, imagery analysts, are charged with intimate acts of witnessing and are embedded in the visual footage of targets. This proximity of the imagery analyst to the target is thought to result in more efficient killing practices. "HB," a drone whistleblower, explains:

> There are people who exist in that world who do imagery analysis only, their primary responsibility is to observe what's going on and fulfill the view of the camera, and to basically reiterate in constant, an ongoing ticker, of explaining, of writing out everything that is happening, every person that shows up, every person that leaves, the amount of children, the amount of women, the amount of men, every color, door, window, so on. (Interview)
>
> Imagery analysts are typically straight out of high school, they're 18, it's their job to look and see if this person's a terrorist or not, and it's their job when they say, "Yes this person is a terrorist," that that person then ... you are actually watching people, and you are not sure, *this is actually very traumatic for the person who realizes that they are deciding between life and death.* And it's particularly hard for the imagery analyst and the sensor

> operator; pilot it doesn't matter, even if the pilot fires the missile, the pilot doesn't see an image. They're given a set of coordinates and they press a button. Just like a jet pilot has no contact with the person; they generally don't have any remorse, they're just fulfilling an order, and if they really believe in the system or the war then, if you're an imagery analyst then you're actually watching people and you know what their lives are like, and doubt creeps into your mind. And it should, because most of the time these people are innocent. … In none of these cases, when these people are killed, they're never actually in the middle of an act, never. They're always at home, or sleeping, etc. The pilot has a difficult task, keeping the plane steady and moving the plane etc., they're not really watching anything, and they're looking at a computer screen. The sensor operator is managing a camera; it's more of a tactile involvement in the act of killing, if they push the button. (PC, Interview)

The imagery analyst,[11] as described here, is a young kid "straight out of high school" who is nonetheless tasked with documenting and analysing "everything" that appears on screen with a degree of certainty and objectivity such as to "decide between life and death." The Bureau of Investigative Journalism's (Fielding-Smith & Black, 2015) report on how the private sector is profiting from the Pentagon's "insatiable demand"[12] for drone war intelligence includes an excerpt from an imagery analyst that further elaborates on the nature of the labour pertinent to imagery analysis. The authors summarize that

> if he [the imagery analyst] thought the images showed someone doing anything suspicious, or holding a weapon, he had to type it into a chat channel seen by the pilots controlling the drone's missiles. Once an observation has been fed in to the chat, he later explained, it's hard to revise it – it influences the whole mindset of the people with their hands on the triggers. "As a screener anything you say is going to be interpreted in the most hostile way," he said, speaking with the careful deliberation of someone used to their words carrying consequences.

Later in the report, titled *Privatized War*, it is revealed that this imagery analyst does not wear a uniform, and he does not work for any branch of the US government (Fielding-Smith & Black, 2015). Instead, he works for one of a cluster of companies that have made money supplying imagery analysts to the US military's war on terror. They claim that "the defense industry's supply of equipment to drone operations is well known, but the private sector's role in providing a workforce has been harder to pin down." Nonetheless the Bureau was able to trace the contracting histories of eight companies that have provided the Pentagon with imagery analysts in the past five years. It is worth citing this portion of the report at length:

> In 2007, defense behemoth SAIC – later rebranded Leidos – was contracted to provide services, including imagery analysis to the Air Force Special Operations Command (Afsoc). A contracting document described SAIC's involvement as "intelligence support to direct combat operations." Its 202 contractors embedded in Afsoc were providing "direct support to targeting" among other functions (in military-speak, targeting can refer to surveillance of people and objects as well as lethal strikes). In a bidding war to renew the deal in 2011, SAIC lost out to a smaller defense firm, MacAulay-Brown … [that was] tasked to "support targeting, information operations, deliberate and crisis action planning, and 24/7/365 operations." The company asked for $60 million to perform these functions over three years. Although Zel Technologies is now the prime contractor, MacAulay-Brown is providing some of the intelligence specialists the contract demands. Indeed, it is not unusual for analysts to simply move from company to company as contracts for the same set of services change hands. They market themselves on recruitment sites with a surreal blend of corporate and military jargon. One boasts of having supported the "kill/capture" of "High Value Targets." Others go in to detail about their expertise in things like establishing a pattern of life and following vehicles. (Fielding-Smith & Black, 2015)

Moreover,

> The Air Force is not the only agency that employs contractor imagery analysts. Intrepid Solutions, a small business based in Reston, Virginia, received an intelligence support contract with the Army's Intelligence and Security Command in 2012, scheduled to run until 2017. In 2012 TransVoyant LLC, a leading player in real-time intelligence and analysis of big data based in Alexandria, Virginia, was awarded a contract with a maximum value set at $49 million to provide full motion video analysts for a US Marine Corps 'exploitation cell' deployed in Afghanistan. Transvoyant had taken over this role from the huge Virginia-based defense company General Dynamics. In 2010, the Army gave a million-dollar contract to a translation company, Worldwide Language Resources, to provide US forces in Afghanistan with "intelligence, surveillance and reconnaissance collection management and imagery analysis support." In the same year, the Special Operations Command awarded an imagery analyst services contract to the firm L-3 Communications, which was to net the company

> $155 million over five years. Defense industry giants BAE Systems and NSA whistleblower Edward Snowden's former employer Booz Allen Hamilton are also involved in the US's ISR effort. BAE Systems describes itself as "the leading provider of full-motion video analytic services to the intelligence community with more than 370 personnel working 24 hours a day." The Bureau has traced some of the activities it carried out through social media profiles of company employees. People identifying themselves as video and imagery analysts for BAE state that they have used real-time and geo-spatial data to support tracking and targeting. A job advert posted on June 10 by BAE gave further insight into the services provided. The posting sought a "Full Motion Video (FMV) Analyst providing direct intelligence support to Overseas Contingency Operations (OCO)" to be "part of a high ops tempo team, embedded in a multi-intelligence fusion watch floor environment." Booz Allen Hamilton has also aided the intelligence exploitation effort for special operations command at Hurlburt Field. Its role included "ongoing and expanding full motion video PED [processing, exploitation, and dissemination] operational intelligence mission," according to transaction records. A recent job ad shows the company is looking for video analysts to join its team "providing direct intelligence support to the Global War on Terror." (Fielding-Smith & Black, 2015)

Finally, this section of the report concludes:

> The hundreds of millions of dollars paid to these companies for imagery analysis represent just a fraction of the private sector's stake in America's global surveillance effort. The Bureau has found billions of dollars of contracts for a range of ISR services. These include the provision of smaller drones, the supply and maintenance of data collection systems, and the communications infrastructure to fly the drones and connect their sensors with analysts across the other side of the world. These contracts have gone to companies including General Dynamics, Northrop Grumman, Ball Aerospace, Boeing, Textron and ITT Corporation. (Fielding-Smith & Black, 2015)

In the air force, contractors are still a minority of the workforce – the anonymous contractor ("John") interviewed in this report estimates that contractors represent around an eighth of the analysts working in support of Special Operations – contractors still "fill the gap to give enough manpower to provide flexibility for the military to do things like take leave" since, "in the military no-one's obligated to respect your time," explained John (Fielding-Smith & Black, 2015). "John argues that taking on even a small number of contractors helps ease the strain on the uniformed force without incurring the expense of pensioned, trained, health-insured employees" (Fielding-Smith & Black, 2015). Furthermore, John and other informants agreed that contractors are highly professional, providing a concentration of expertise:

> By the time an airman has built up enough experience to be competent at the job it's usually time to change their duty location. Age also has a lot to do with the professionalism of contractors. Most contractors are at the youngest mid to late 20s, whereas Airmen are fresh out of high school, said one analyst. "As a full motion video (FMV) analyst, you cannot identify something unless you've seen it before." (Fielding-Smith & Black, 2015)

Private contractors are exposed to normal conditions that characterize ordinary business practices: "You can get fired," notes John, claiming that knowledge as a motivational factor for contractors, as companies' bids and their performance are heavily scrutinized (Fielding-Smith & Black, 2015). A contract may include a built-in incentive fee. There is a clear free-market logic operating here: In theory, the best staff and the most meticulously managed company ought to successfully acquire contracts, while the details and implications of those contracts remain obscure.[13]

As previously discussed, the deaths of private military/security corporation (PMC) personnel are not made public in the ways that the deaths of American soldiers are. Hence, "the turn to the military contractor represents an attempt by officials to designate, by law and policy, a class of persons whose deaths will be banal and insignificant to a national audience. Their existence and relation to the body politic is one of contract, not sacrifice" (Taussig-Rubbo, 2009, p. 85). The revolution in military affairs, coupled with, as detailed in the bureau's report, the "insatiable" demand for ISR intermediaries gives credibility to PMCs in such a way that allows them to render their practices non-political through its claims to be practising healthy economics, not politics. Although contractors working in intelligence analysis are certainly unlikely to be killed or injured on the job their inclusion within a Distributed Common Ground System[14] reflects how the neoliberal demand for both flexible citizenship and free-market capitalism – synthesized in the act of privatization – has deeply penetrated the military sphere, where contractors solve the problem of personnel shortages, a deficiency in expertise, and, perhaps most importantly, are not subject to the usual benefits that are afforded to military personnel as set out by the Department of Veteran's Affairs.

In the case of imagery analysts, and all workers within the drone labour hierarchy, it is not their deaths that will be rendered banal, because they are not exposed to such risk. Rather, their mental traumas may potentially be excluded from a rights-based, collective and/or nationalized narrative. Instead, such suffering would be managed by reducing work hours and hiring more people to share the burden of increasing combat air patrols (CAPs) through even more privatization and even more technological solutions.[15] In early 2015, "the Air Force operated 65 combat air patrols, or up to 260 drones patrolling war zones around the world, but was under pressure from the Pentagon to expand the number of drones in the air to 360. But pilots, overwhelmed by the complexity of their work and the grueling schedule, began to vote with their feet and quit" (Thompson, 2015). This created a critical shortage of pilots (which, as the *Air Force Times* reports (Losey, 2016), has seen the air force respond by offering incentive packages and retention bonuses of upwards of $175,000 if they agree to a five-year active-duty service commitment) and thus invited privatization to fill the gap. Meanwhile, the air force plans to add 65 new Reapers to its fleet, doubling the number of pilots, as well as facilitating the development of new drone operations centres at air force bases around the United States (Losey, 2016). Two companies have been awarded contracts from the air force to provide drone support services: The first, Aviation Unmanned, is a small, veteran-owned, and operated business. The second is General Atomics, a large, California-based military contractor that manufactures both the Predator and Reaper drones (although companies such as Booz Allen Hamilton, General Dynamics, and SAIC have previously held contracts to analyse surveillance data acquired by drones operating over war zones; Losey, 2016). Notably, Major Keric Clanahan, legal advisor to the US Special Forces, warns that "it is imperative that contractors not get too close to the tip of the spear" (Clanahan, quoted in Thompson, 2015), suggesting the sacredness of killing is the realm of citizen-soldiers who act in the name of the sovereign authority.

The privatization of war, after all, is not simply an economic manoeuvre but also a social and deeply political one. It is worth quoting an excerpt from Heather Linebaugh, a former imagery analyst featured in the film *National Bird*. In an article for the United Kingdom's *Guardian* newspaper (Linebaugh, 2013), she asks politicians who defend the US Predator and Reaper drone programs, "How many soldiers have you seen die on the side of the road in Afghanistan because our ever-so-accurate UAVs were unable to detect an IED that awaited their convoy?"

> What the public needs to understand is that the video provided by the drone is not usually clear enough to detect someone carrying a weapon, even on a crystal-clear day with limited cloud and perfect light. This

> makes it incredibly difficult for the best analysts to identify if someone has weapons for sure.

Not all people involved in drone warfare are hesitant, self-reflexive, and full of self-doubt. An interview with "JA" attests to this. He joined the army as a senior in high school in 2003 and served on active duty as a Fire finder radar repairer and operator from 2003 to 2009. He was stationed in Fort Campbell, Kentucky, with the 2nd Field Artillery Detachment, 101st Airborne Division and was deployed to Iraq from 2005 to 2006. He did a year in South Korea with the 2nd Infantry Division from 2006–2007 and then was stationed with 3–16 Field Artillery, 2nd Brigade, 4th Infantry Division in Fort Carson, Colorado. He was honourably discharged in early 2009. He claims that

> the drone operator is doing all of this observation, meaning that they could be watching a target for hours or for several days in a row before blowing that person up. The operator may see the target going about day-to-day activities or around his family, so all that time watching can lead to empathy with the person being observed. Once it is time to launch a missile at the individual, drone operators may feel guilt or regret about what they have done. In contrast to the above, I think that in some respects remote/electronic warfare could make it easier for soldiers to pull the trigger. Some drones used overseas are operated by pilots within the continental US, meaning that the pilot is removed from the conflict and can go about his/her normal day in between killing people remotely. That disconnect probably makes it easier to pull the trigger. (JA, Interview)

The dual nature of drone warfare – as both generating/rejecting empathy – as it manifests in the subjectivity of the operator is perhaps too dichotomous in its rendering. What matters most is how a style of violence underpinned by intimacy-in-distance splits the agent of violence at the level of the body: the act of killing, with no exposure, no vulnerability, no fight-or-flight response inside the killer(s). Even those on the ground are often miles away from the perceived enemy such that the risk of physical entanglement is minimal. Yet, given the nature of the work itself, there is plenty of time to entertain, as Linebaugh attests to in the next excerpt, "haunting memories," and the guilt of always being "a little unsure." The speed of killing – from the determination that a kill is imminent to the strike itself – is matched only by the sluggishness of affectivity; the emotional making sense of the event:

> One example comes to mind: "The feed is so pixilated, what if it's a shovel, and not a weapon?" I felt this confusion constantly, as did my fellow UAV

> analysts. We always wonder if we killed the right people, if we endangered the wrong people, if we destroyed an innocent civilian's life all because of a bad image or angle.
>
> It's also important for the public to grasp that there are human beings operating and analyzing intelligence from these UAVs. I know because I was one of them, and nothing can prepare you for an almost daily routine of flying combat aerial surveillance missions over a war zone. … Proponents claim that troops who do this kind of work are not affected by observing this combat because they are never directly in danger physically.
>
> But here's the thing: I may not have been on the ground in Afghanistan, but I watched parts of the conflict in great detail on a screen for days on end. I know the feeling you experience when you see someone die. Horrifying barely covers it. And when you are exposed to it over and over again it becomes like a small video, embedded in your head, forever on repeat, causing psychological pain and suffering that many people will hopefully never experience. [We] are victim to not only the haunting memories of this work that they will carry with them, but also the guilt of always being a little unsure of how accurate their confirmations of weapons or identification of hostile individuals were.

Being "unsure" seems integral to the act of signature strikes, which target people who are not identified but merely display a certain "pattern of life." CIA members, for example, do not personally identify these targets, but rather, "they exist as digital profiles across a network of technologies, algorithmic calculations, and spreadsheets" (Shaw, 2013). Indeed, what Shaw (2013) names a topological spatial power means a drone's victim never hears the missile that kills him (Rhode, quoted in Benjamin, 2012, p. 149). Nevertheless, in the face of mounting pressure around the legality of drone strikes as per international humanitarian law, former president Barak Obama, having barely admitted to a drone program, said in May 2013, "dozens of highly skilled al-Qaida commanders, trainers, bomb makers and operatives have been taken off the battlefield. Plots have been disrupted. … Simply put, *these strikes have saved lives*." The notion that pre-emptive strikes save lives is a statement that assumes an imagined schematic of risk, which requires the audience to buy into the biopolitics of civilized Western life (which ought to be protected) versus uncivilized life.

The idea that drone strikes are ultimately effective because they save lives is abundant in my interview data. In the first two passages the follow, drone technology is credited with saving lives, while in the second two passages, it is credited with saving lives as well as reducing collateral damage. It is notable how *life* and *collateral damage* emerge not as synonyms but as antonyms, positing the proper subject of life, that which deserves protection, as the Western/ North American citizen-soldier. The implication of this distinction is that incorporating and adopting sophisticated technology is essential to the reduction in loss of Western and/or North American life, which also positively correlates with the minimization of collateral damage – this being a desirable by-product of drones but nevertheless not the primary priority:

> Put it this way, you blow a drone out of the sky, they shoot it down, so what? In less than in a week we'll have a new one in Afghanistan, flying again. We'll rush one on a plane; very quickly they can have another drone there. It's a major issue if someone gets killed while on a reconnaissance mission. So, the cost is one thing, but if you blow ten drones out of the sky no one will be upset except the finance guys, and *if you kill ten guys that will be on the front page* of the Globe and Mail. So cost avoidance, yes, but I think life-saving, *there's a strong argument that drones save lives,* I would suggest. (GB, Interview)
>
> More than half of the pilots there are unmanned pilots; the air force has really embraced that particular technology. Look at how many predators and global hawks are out there and what they're doing, and even on the small side, all the services are embracing the smaller UASs because *they're so capable of saving American lives.* (TW, Interview)
>
> Well, the question becomes the missions of the past are being redefined for missions of the future, so missions of the future will not need soldiers – "boots on the ground"– in foreign wars, as compared to [the] Korean War, or [the] Second World War, or Vietnam for that matter, or even the first Iraq War. Those kinds of troop movements are going to be less frequent, at least based on the trends that I see. Something will be lost. Human intelligence will be far more critical. There are all kinds of futuristic ways people imagine what war will be about. Our accuracy of killing has become much better, our collateral damage has become much less. The electronic sophistication, you can blow up a house, and the houses on either side of that are going to be untouched. It's all part of electronics, whether manned or unmanned, and so conducting war from a distance ... and again that has been going on, it's just that what we're seeing now is greater sophistication because of technology, but the trend in terms of what is actually being accomplished is not much different it's just that the execution is quite different because of the novelty of technology, and the expense of feeding soldiers. (GS, Interview)

Indeed, when troops do move, it is a few select, highly trained individuals (usually special operations forces). They operate at night to minimize detection and remain largely invisible. Another interviewee verifies the appeal of special operations teams. He, like the previous interviewee, confirms the value of leaner forces integral to today's wars:

> I was then indoctrinated into the Joint Special Operations Command (JSOC) basically they are of their own … if you think in military terms, basically if you can imagine them as a separate branch of the military, but collectively all of the best of the services, but they're relatively small, there's no fat on them, they don't carry any dead weight, but at times when missions ramp up, especially now with the threat of terrorism popping up in several different countries all over the place, they need additional support. When they need it, they tend to get it, and they get priority over most people and the positions have to be filled, that was basically what I was filling, was the need for them to have support in intelligence to come and be a part of their world for a little while and then I'd go back to doing my own thing. When dealing with JSOC they considered themselves to work more effectively, to cut through red tape faster, to get around issues that prevent most conventional forces from basically being as effective as they could be. So basically JSOC is very streamlined process, organization, for doing special operations and they worked at what they say, it's like hitting the treadmill at 100 miles an hour, they don't have time for people who lag behind, or people who don't understand, or people who can't give a straight answer immediately, so they have a certain level of professionalism that is somewhat in a way admirable and refreshing compared to the politics and bureaucracy and sluggishness and insufficiency of the military in general, I mean when I came into the world of JSOC, there was this kind of I guess, call it appreciation, for what they were doing. It was my responsibility to train any new person that arrived in the country, because even though you get trained before you get there, it's still not real-world experience, you need to have some sort of training to make sure you aren't going to be making any major mistakes because it's a very important thing that you're doing, people's lives are on the line, not just our own, but because your intelligence leads to what could be sending a strike team out but also the lives of potentially innocent Afghans being caught in the crossfire. I took the position very seriously, but the position itself did not require an immense amount of effort, just simply required endurance and attention, you had to be paying attention, and you had to be ready to respond any time you were called upon. Sometimes you would go the whole twelve hours and absolutely nothing would happen, and sometimes you would be working three missions at a time for that entire twelve hours. There were people who worked night shifts, and there were people who just worked day shifts. Typically, people who worked day shifts were in the process of just doing general monitoring to conduct

> surveillance and gather intelligence on high-value targets and people who work at night were usually doing support for nighttime missions, nighttime raids. Basically, you know it was very common to send out a strike team, you know, a few strike teams go out every night and they try to capture the individuals that they've been monitoring for however long, if enough intelligence has been gathered to lead to them believing that they can actually capture the person. (DH, Interview)

The use of special operations forces, described in opposition to the "sluggishness and inefficiency" of the military have a direct line to the president and follow their own rules of engagement. Recall the last line in the final excerpt. DH explains that it is common to send strike teams at night to capture people that have been under surveillance "if enough intelligence has been gathered to lead to them believing that they can actually capture the person." Given interview data cited previously from legal experts regarding the shift from a capture-to-kill mentality, this point should cause pause. It remains doubtful that the United States would consistently aim to generate the necessary intelligence to capture non-US citizens, especially as disembodied war seeks to avoid embodied encounters.

Summary

Nearing the end, we find ourselves back to the beginning. Recall the botched night raid in Yemen that this book began with, where a "grand display" amounted to the death of US Navy SEAL Ryan Owens. Bill Owens, Ryan's father, did not agree with the dominant political discourse that his son's death had any sacrificial value. He refused to meet with then President Trump, who, in contrast, wanted to paint this loss as a death worthy of sacrificial language, as I have argued, as way to normalize and justify the raid and the resulting injuries and death. As the interview data cited in this chapter show, the increasing use of drones has not come out of thin air, but is instead shaped by political and social conditions, which give rise to a particular style of war – a direct response to something. It is that something we ought to be most curious about as we think about future war. This style of war disrupts a taken-for-granted sacrificial politics. In doing away with not just with the citizen soldier archetype, but importantly its associated values, we can see a wider approach to war that is actively shedding its legacy in the spirit of French revolutionary and later "democratic" warfare. Put simply, war and the People are on unstable grounds. Perhaps this has always been true to some degree. Whether this is a "good" or "bad" thing does not matter; to what extent the French Revolution was about true democracy in practice (it was not) does not matter for our purposes. What matters is the acknowledgment that we have inherited norms and traditions around war in liberal democracies, and when privatization combines with drones, some important practices and values might be lost. Conceptually, sacrifice gives us a portal into this loss – the subject of the next, concluding chapter.

Chapter 6

Antisocial War

Every act of identifying an enemy is fraught with risk, for if the populace fails to see that person or group as the enemy, it will see only murder, not sacrifice.

– Kahn (2011, p. 156)

The fact that slaughter is a horrifying spectacle must make us take war more seriously and not provide an excuse for gradually blunting our swords in the name of humanity.

– Clausewitz (1976, p. 260)

The Future of War

One of the important things that might be lost in the context of disembodied, privatized drone wars is society's ability and desire to remember war. As emerging technology permits increasingly disembodied forms of warfare (displacing risk as an element of combat) the roles of humans in battle are likely to continue to become more invisible, fragmented, and depoliticized. Taking disembodied war to its extreme, where would the space for war memory be located, absent a wounded body, and absent witnesses who can account for violence from beginning to end? While "automated warriors" might not be exposed to physical risk, they can endure moral injury (Asaro, 2013; Enemark, 2019). That these are mental and not physical wounds, and that the labour of automated warriors can blur with routine civilian digital work, displaces our ability as a society to associate new forms of combat with trauma. Because mental wounds are viewed as "less than" physical wounds, they do not "count" in the register of trauma that has historically informed war memory (Baggiarini & Rupka, 2020). Without trauma located in the physical body the opportunity to remember disembodied combat is eclipsed by the qualities of increasingly "algorithmic" war (Amoore, 2020; Aradau and Blanke, 2022; Crawford, 2021) – defined by invisibility, fragmentation, and depoliticization (Baggiarini, 2024) – thereby inducing the displacement of humans (physically and cognitively) and thereby a form of collective forgetting.

Forgetting may help some cope with war's devastating psychological effects, but as a national society, remembering war is key to (a) binding liberal democratic society with the military which acts on its behalf, (b) bridging an understanding of the harms and impacts of war across generations, and (c) building a basis by which to prevent and mitigate against future wars. Without shared memories, the violence of war may move from its legal and ethical status as exceptional, towards something much more banal. To ensure war remains an exception rather than the rule, historical memory (although contested and not straightforward) can and should guide future technologically mediated military organization and planning in line with the ethics and law of warfare. While conventional twentieth-century wars have historically seen an influx of memory studies (the Holocaust being the prime example – see Winter, 2006), unconventional and contemporary or limited wars must also be remembered to maintain the core principles of liberal democracy. Without sufficient modes of witnessing and remembering, we risk increasing forms of antisocial war. This is a problem for liberal democracies, which need remembrance, and some understanding of sacrificial language therein, to not only "know" and remember war but to prevent future war as well.

Is the future of war to be found in a return to the past? On the surface, drone technology not only appears disconnected from the past but also reflects a desire to overcome some past constraints in war around the assumed requirement that combat involves bodies and embodied encounters on a battlefield. The highly influential military strategist Carl von Clausewitz (2010) viewed war as combat, and combat as slaughter. Integral to war for him was risk, injury and death to soldiers: Through these injuries that amount to slaughter, and thus a horrifying spectacle, we can take war seriously and we can *know war*. Drones theoretically might help us get around some aspects or tactical difficulties relating to the slaughter part, in addition to making war more predictable and uneven, in line with the twin mandates of virtuous war and casualty aversion. But in doing so, the question of what remains which would allow us to take war seriously – the question of how we take war seriously – remains unanswered.

Yet some will say that the history of warfare can be understood as a continuing practice of how best to inflict lethal harm while protecting one's own troops. To this end, some may argue that casualty aversion has always been a consideration for military strategy. They would point to a series of technological improvements- from spears, boomerangs, and bows and arrows to gunpowder, hand cannons, and other small arms to rockets, shells, machine guns, and then tanks and nuclear bombs, to name some key developments – as a sign of casualty aversion's broader historical rationale and relevance, and its realization through technological means (consider the notion of "distanciation," the importance of putting greater distance between one's troops and one's enemy, or Paul Virilio's concept of "dromodology," an argument that echoes that of distanciation, but with a focus on how acceleration and speed, made possible through technology, now trumps geopolitics), all of which precedes the Vietnam moment. However, this is (a) techno-fetishistic, and deterministic, and ignores human intentions and (b) misses how the Vietnam and post-Vietnam era produced critical effects around the policing of wartime suffering and memorialization. This moment did not just inspire casualty aversion as a political universal ideal (uniting both American liberals and Republicans in this common interest), but it troubled the sacrificial idioms associated with citizen-soldier bodies and their relation to war-making. The "Vietnam moment" then is about more than how it formalized an ideology of casualty aversion. It signals sociopolitical ruptures and congruent effects around hyper-individualization, neoliberal flexibility, and the decline of the nation-state that are relevant for thinking about military sacrifice specifically. Regarding the first point, the danger is in attributing technology with its own internal essence; treating technology as if it bears its own rationality, distinct from the humans who employ it. There is an assumed linear historical progression that becomes tethered to the understanding of technological improvement. For example, writing on human-machine interface, DeLanda (1991) claims "the events on a computer screen may also become elements in a strategy to set humans out of the loop, to shorten the chain of command. This seems to be the direction where machine vision and machine translation are going" (p. 193). Aside from the obvious criticism that "machine vision" and "machine translation" *simply cannot be* without a human interlocutor, the question of how intelligent machines displace humans – by what social or political mechanisms – remains undertheorized yet is critical. Large drones, in fact, rely on dozens if not hundreds of bodies in their operation, but it is the political significance associated with their bodily presence or absence that signals the most profound form of displacement. On the second point, if the nation-state bears some residue of sacredness and if there are citizen-soldiers to embody this, then sacrifice will continue to be evoked as both an important theme, and in practice. However, since Vietnam, the United States has clearly moved towards privatizing many aspects of warfare (consider how the Bush administration sought to minimize the circulation of images of flag-draped coffins) such that wartime experiences of loss and suffering become disassociated with the nation-state's right to violence (consider that no new memorials have been erected stateside for those who died serving in the Iraq and Afghanistan wars). Should 'we' in the 'West' remember the War on Terror? If so, for what purpose?

Drones, based on the philosophy of privatization, signal the end of the notion that one must be prepared to both kill in die in war (Asad, 2007). In doing so, drones are thought to offer greater possibilities for precision and thus more accurate and humane forms of future war. But the answer to what future war might be is not found in greater humanism but perhaps requires a backward-looking view of the past. Consider Clausewitz again: It is *because* combat is slaughter that we can take war seriously and presumably prevent it. Absent visible, imminent, and shared risk or injury in war can we not take war seriously? Does this mean the solution is that we should have more death and suffering in war or that we should return to a bloody past rooted in empire, hyper-nationalist zeal, and or conquest and domination so that we can then say something meaningful about ourselves, about society? Of course not. Yet, how will we be able to know, remember, and prevent future wars if our end goal is indeed virtuous war, to avoid harm to soldiers in war full stop, because it is through the citizen soldier as an archetype and associated values that we have *historically* been able to understand and convey the myriad costs of war. Absent the vulnerable body of the citizen-soldier, and all the social and political values wrapped up in the archetype, we should ask where it is – if not the body – that we can locate the political and ethical qualities of war. Or will machines, in all their perfectibility, take up our ethical aspirations, as Elke Schwarz (2019) cautions? While rhetoric links citizenship to sacrifice (Kahn, 2011, p. 80) "for many today, the link of citizenship to identity and identity to sacrifice seems a memory – a remnant – of a less cosmopolitan, and more violent past" (p. 99). To be sure, military sacrifice is fundamentally about violence, and war is about violence. Without sacrifice, can we understand the violence of war? This book has shown how drones and military privatization together provide us with a framework to understand the breakdown between citizen-soldiers and sacrificial cults and idioms.

Yet, drones are becoming more complex and capable through advancements in artificial intelligence, particularly developments in machine learning, reflecting a synthesis between old and new. On one hand, drones are being used in new and unprecedented ways. On the other hand, they signal a turn to the past, where wars were limited, dynastic, and private affairs between kings and their highly skilled mercenary armies. In these wars, civilians were not directly involved. It would only be at the start of the twentieth century with the Russo-Japanese War of 1904 and later the Great War when civilians and citizens, with the help of advancing communication and media technologies, began to take notice and or become directly involved in either soldiering, supporting, or supplying war. The point is that we ought not be distracted by the technology of current wars but instead ask what this technology tells us about the meaning and place of bodies in war, and about the changing quality and character of war as it relates to 'the social'. Sacrifice gives us a clue into this. But concepts of non-sacrificial or post-sacrificial war are not well supported, and seem incomplete as a conclusion to the problem of governing military sacrifice. As I conclude here in this final chapter, disembodied wars still involve material sacrifice and still involve the politics of sacrifice, as seen in attempts to sometimes sell wartime death and injury as sacrifice. In looking at the meaning and place of bodies in war, and the meaning and place of drones in war, what they together point to is a marked shift, not to post-sacrificial war exactly. Instead, they point to war in some instances becoming *antisocial.*

Antisocial war

Antisocial war needs to deploy a governing logic to be able to both demand and disavow sacrifice, but it is driven by a desire to avoid embodied encounters as conventional troop deployments become more politically and logistically hazardous. Through the concept of antisocial war, which does not take military sacrifice as given or universal, we can become more curious about moments when sacrifice appears to be silenced, in addition to moments when it is summoned and celebrated. In doing so, we can ask questions about war's ontological status. Moreover, antisocial war is illustrated through the disconnect between a national society and its war-making apparatus. I have argued that this disconnect can be seen in the breakdown of citizen-soldiering's link to sacrificial cults and idioms. Likewise, as LL claims,

> I believe I lost some of my humanity while I was in the drone program. All of the words that they used were clean, sanitary … what was really going on was that people were getting killed and dismembered in areas where we may or may not have been at war. And although I wasn't involved

> directly with any of the strikes, working on a distributed system, I think everybody should carry some of the responsibility. I mean I think the people that live here should carry some of the responsibility of a war that is being carried out in their name. And that's not happening. (ML, Interview)

Some of ML's humanity was lost in the drone program. This implies that war, and those who do the work of war across strategic, operational, and tactical levels, ought to have some concept of humanity, some concept of self, community, purpose, and embodiment, to render it meaningful. For ML, drone warfare lacks meaning and justification, even at is packaged in sanitary and virtuous language. Meanwhile, what is made invisible are the undeclared conflict zones and the violence, both embodied and disembodied, that occur there. Embodied violence has been integral to notions of war and combat, culminating in the modern period when the People claimed sovereign presence and capacity for violence. The global scope of militarized violence has meant that states have had to look beyond their national borders for not only labour but also justifications for continued violence. The war on terror troubled the temporal and spatial limits to warfare, disrupting legal categorizations and precedents for war and violence therein. The implication of this style of warfare has profound effects as far as how it transforms social relations. For LL, drone warfare is happening while people seem unable to take responsibility for wars carried out in their name. This individualization and avoidance of witnessing and responsibility is reflected in the statement of another drone critic and activist. They told me:

> I remember when I was a kid my parents would go down to the post office and we would get ration cards. Everybody had to paint the top half of their headlights black so when the Germans flew over so they couldn't see the cars. It was a way to make everybody feel like we're in this together. You're not going to get to buy everything you want because some of this goes into making tires for airplanes, so people realized we're in it together. This is not the situation now. (NM, Interview)

Drone warfare exploits and accelerates multiple forms of inequality, and exacerbates an inherent tension between "the people" and the state, revealing its need to be made and remade over time. This interviewee situates their learning about national community, defined as social relations, through an experiential lens of warfare, and the small sacrifices made therein ("you're not going to buy everything that you want"). Although we might be critical of the link between identity and violent nationalized imaginaries, asking rightly if the latter must necessarily inform the former, this testimony nevertheless suggests that the value of sacrifice in war is to be found in its communalizing potential ("we're in this together"). This communalizing potential can be problematic, and we can point to many historical and contemporary examples of this justifying exclusionary, hateful, and violent practices. But, for better or for worse, for state-sanctioned war to be legitimate, we have come to believe that it must reflect the legal, political, and social norms that organize a community's consent in relation to violence.

Like NM earlier, "JA" draws on the theme of sacrifice in his description of how he was unwillingly subject to the military's practice of "stop loss," which is used to involuntarily extend a service member's contract at the end of their enlistment terms:

> I was also opposed to stop loss because I saw it as a back-door draft and a way to continue using our overextended military to fight multiple wars *without asking the American people to sacrifice much.* Instead of risking a political backlash by instituting a draft, the government was involuntarily forcing those of us who volunteered for service to stay for even more time. Instead of asking the American people to share in the sacrifice, those in the military (who often come from poorer and minority backgrounds) were deployed repeatedly. (Interview)

Few other events other than war have a way of bringing into view the link be sovereign power and sacrifice. The idea that the burden of sacrifice be shared or at least more distributed seems to be lost. According to this interviewee, a notion of shared sacrifice "is not the situation now" and "instead of asking American people to share in the sacrifice," we have instead the continuing expectation that war requires increasingly fewer bodies; that some bodies are necessary, but nations and populations are not. The military's use of stop loss is an egregious example of how those who had already served were compelled to serve repeatedly since the overextended military of the post-9/11 era could not possibly make the political case for a draft or the requesting of additional troops. Drone violence, extending from casualty aversion to a technological response to military/security privatization has perhaps unsurprisingly become the technological "face" of future warfare, but it is paradoxically rooted in a distant, pre-revolutionary past, one that we, through a notion of history as linear and progressive, problematically assume has been long forgotten.

I have argued that there is a paradox in the practice of military sacrifice, which is revealed and exacerbated by the neoliberal restructuring of capitalism's link to war. Governing is the response to this paradox, illustrated by shifting ideas of how citizen bodies ought to *be* in war and what meaning can be gleaned from the relative presence or absence of bodies. In contrast to the dominant theological/philosophical literature, which I claimed runs the risk of essentializing, de-historicizing, and depoliticizing sacrifice, I offered a critical political-sociological approach, which is concerned with the positioning of bodies in the organization of violence and the changing meaning of the citizen-soldier archetype. I examined how sacrifice becomes depoliticized, and under what conditions it includes or rejects the body and/or technology. I argued that political discourse on sacrifice reveals this precariousness of the citizen-soldier's relation to sacrifice, and because of this precariousness, attempts to manage the contradictions of the neoliberal restructuring of capitalism's link to war are brought into sharp relief. In the Canadian context, the examples I cited show how sacrifice can be deployed in both a celebratory manner (re-nationalizing and re-militarizing identity) and the democratic pulse of the sovereign nation is explicitly linked to the memory of honourable past sacrifices. In the US case, I argued that sacrifice is deployed in a hesitant and cursory manner. It is un-problematically regarded as essential to victorious warfare but nevertheless presented as ideally no longer required in downsized and high-technology war, where bodies are treated as liabilities rather than assets. The problem here is that bodies in war are carriers of political values and ideals, and their deployment in war seems to signal, ideally, that wars are just and reflect the authorization of the People. Without bodies, politics do not disappear but are reconfigured.

Themes of justice and justification (casus belli) permeate the idea of military sacrifice today. Like politics, sacrifice is not being eclipsed; to be sure, it is reconfigured too as it is both demanded and disavowed. But sacrifice must not only be held in reserve; it is a reserve that must be visible to the people. Sacrifice is increasingly contentious; the value of the sacrificial subject now is so great that only in extremis is it an act that must be made. Military sacrifice must appear alongside a schematic of justice. If the people regard a war as just, if citizen-soldiers exist to bear at least a residue of the nation-state's sacredness, then a sacrificial impetus is, in theory, more effortlessly applied by the state and its military apparatus. Considering how this sacrificial subject is being placed at the edge of modern politics – so fraught in today's liberal societies, whose ever-elusive promise of peace remains the context for the conduct of international politics – the question of how to legitimately apply violence remains. The solution appears, aside from withholding violence altogether, to conduct it in the margins, its visibility and invisibility being guided precisely by a politics of violence that views sacrifice as that which must be readily demanded as well as disavowed. However, the presentation of a "just war" is not in and of itself sufficient to motivate cohorts willing to sacrifice. Not simply influenced by the "Vietnam moment" but also growing individualization, the reconfiguration of the nation-state, and flexible citizenship, together these processes have suggested the existence of political forms that complicate the simple link between the re-presentation of a just war and its implied sacrificial logic. Even the language of "just war" appears, following the arguments of this book, somewhat empty. I have claimed that, rather than "war," in a traditional sense, we have witnessed in recent years a series of violent encounters, revealing not violence's renewed formalization but rather its increasing *banality* wrought by its temporal and spatial expansion.

Furthermore, while military privatization is the sociopolitical expression of this reconfiguring of citizen-soldiers' link to sacrificial military violence, drones are the congruent technological expression. While both military privatization and drones can be said to have harmonizing effects on the problematic of sacrifice, I argue that their relation runs much deeper than this fact alone, for both drones and privatization are symptomatic of the same philosophy of violence: Both are about the desire to enact greater forms of distance – political, physical, affective, cognitive – between the nation-state's violence and its citizens. Private military/security corporations perpetuate this distancing by turning state violence into an economic problem instead of a sociopolitical problem. While there is a marked moral and physical distancing in privatized fighting, there is also a political distancing. When contractors suffer death or injury, states are not responsible for confronting the consequences or for providing care or benefits, because the relationship between the two is economic rather than social. Military privatization is part of an attempt, quite simply, to make war private. Drones are the technological expression of this, insofar as they are machines with sensory and lethal capabilities that have come to be because of the development of expertise integral to military privatization. Privatization is an economic manoeuvre, in addition to a sociopolitical one. Drones, as we know them today, could not exist absent their foundation in military privatization, on one hand, and casualty aversion, on the other.

Military strategy today seeks to minimize risk posed to certain bodies as exemplified in the ideology of casualty aversion. To minimize risk is to avoid the scene of sacrifice while eclipsing violent embodied encounters. Because this style of violence avoids the scene of sacrifice through the rejection of embodied encounters, drone warfare could be theorized as symptomatic of a desire for extra-sacrificial war, or as a non-sacrificial style of violence. Yet, although drone warfare facilitates a non-sacrificial style of violence, it does not mean that some categories from sacrificial systems past do not remain in cultural circulation, dislodged, however, from a totalizing sacrificial system. To be sure, we have sacrificial fragments, absent the whole. Sacrificial ideals are indeed still resuscitated through premodern, monarchical versions of sovereignty, sometimes with the view to explicitly remembering war and past sacrifices therein. But these sentiments can ring uncritical and hollow, occurring in disharmonious tension with the contemporary reality of wartime violence today which seems adamantly post-sacrificial. Sacrifice is oftentimes conceptually incoherent within the political landscape of the democratic sovereign. As a result, to gain credibility, sacrificial discourse must invoke the symbolism of premodern sovereignty to maintain the nation-state's grip on legitimate violence. Even then, as the epigraph by Kahn (2011) states, "every act of identifying an enemy is fraught with risk, for if the populace fails to see that person or group as the enemy, it will see only murder, not sacrifice" (p. 156). Many of those killed or injured during drone strikes edge dangerously close to a murderous plot as opposed to a sacrificial one, while the body of the perpetrator – the author of sovereign violence – remains largely invisible – the trace of the drone, its visibility, is apparent only in the moment of the strike (Rupka and Baggiarini, 2018).

In this dynamic act of erasure, what is being concealed or dislocated is the origin of sovereign power, understood as that power that is directly tied to the author(s) of state violence. Whereas sovereign power in war zones was traditionally expressed through the embodied and violent encounter characterized by twentieth-century notions of a spatially contained and publicized battlefield, featuring the prominence of the national flag, the uniform, and relatedly foregrounding violence as that which occurs in the name of the people and their nation, today's drone wars circumnavigate the "problem" of sovereign power's relation to biopolitics. Namely, the problem is about how to reconcile liberal democracy's promise of peace with expanding forms of violence. To reconcile this, we see the outright denial of the encounter, combined with the explosion of spatial and temporal limits to violence.

The practice of erasure also figures into how sacrifice has been displaced along three interrelated axes: publicity, surplus, and embodiment. I gestured to the idea that political rhetoric upholds a relation between citizenship and sacrifice in the context of a rapidly changing style of violence where the intensification of privatized and disembodied social relations rather than their redistribution is evident: Here, the speed at which violence is designed and applied is held back by the ideological, citizenship-based commitments to the body of the citizen-soldier and, by extension, the People. Rather than war, in its traditional form, we have targeted, fragmented, and isolated acts of violence, where sovereign power is no longer maintained by or reflected in the body of the citizen-soldier. Considering this rejection of a national audience I argued that sacrifice is disrupted along three interrelated themes or axes, which all hinge on the sovereign and symbolic power associated with the archetype of the citizen-soldier: publicity/privatization, surplus/rejection of surplus, and embodiment/disembodiment. Drone violence is privatized, depoliticized, and swiftly cleansed of its visible aspects – the spatial logic of the battlefield that informed twentieth-century warfare is eclipsed – and as such, history has its traumatic wound erased (Rupka & Baggiarini, 2018). This implies the end of a specific temporal logic, whereby the writing of contemporary history now occurs minus the traditional forms of sacrifice and embodied violence. Structures of meaning, pertaining to citizenship and the nation-state, are removed from the theatre of war.

Drone warfare attempts to transcend the sociopolitical and moral limits of the body, thereby eclipsing the injury of citizen-soldiers on "our" side. This problematically destabilizes traditional means of commemorating wartime suffering – and therefore dramatically limits how wartime state violence can be acknowledged or recognized. Traditionally, sacrificial language has offered a means by which to recognize, know, and commemorate wartime violence. Yet the combination of military privatization – the outsourcing and technologizing of war-making – destabilizes sacrificial idioms, such that the citizen-soldier is no longer capable of binding the nation-state to a credible use of violence. Yet drone violence gains credibility through a dialect of visibility (the battlefield) and invisibility (the violence) by rendering that which was previously unseen visible – casting excess light on the battlefield to break up the fog of war. As such, it claims to save and protect those lives that matter most: American and allied soldiers. However, the political and social landscape in which drone warfare operates and gains credibility, through visibility to make live, depends on the invisibilization of the politics of emotional trauma, the scene of violence, and the identification of the witness.

Summary

This book illuminates how the acquisition of the discursive apparatus of sacrifice, its proliferation in language and popular culture, simultaneously makes claims about the biopolitics at stake in the question of citizenship. Citizenship and sacrifice were tied together during nationalization projects, and this relationship is now weakened due to denationalization and privatization. Flexible citizenship and the private contracting of soldiery capture a shift relative to the sacrificial logic that is bound up with the archetype of the soldier-citizen. If private contractors cannot make any claims of legitimate sacrifice, then this begs the question of why the state might desire the "end of sacrifice." If state soldiers continue to become an increasing burden on the state, because of the political toll of unpopular wars or the financial costs of lifetime support of veterans, the inclusion of private contractors and drones provides a much-needed respite for the state. When war-making is outsourced, it is subsequently rendered invisible, thus challenging the belief that the legitimacy of war depends on a national public validating it. The end of sacrifice would make the death of soldiers profane.

Together, military privatization and drones reveal the tenuousness of the citizen-soldier's link to militarized sacrificial idioms and cults. This decay of the citizen-soldier's place in the modern concept of the People, and its related erosion within the nation-state's articulation of legitimate violence, challenges two significant prevailing assumptions of our time as pertains to the ontological status of warfare: that militarized violence (a) is always-already sacrificial and (b) links trauma to a certain kind of aesthetics of bodily suffering and a grammar of transcendence for the nation. If violence becomes increasingly removed from the sacrifice of the citizen-soldier body and if violence is increasingly omnipresent and panoptic, then so, too, is the event of the violent encounter effectively dislodged from a politics of memory, trauma, and therefore an affective schematic of sacrifice. Trauma and sacrifice become isolated and individualized, so a national narrative cannot lock into it and, in fact, has no compelling reason to. This creates an even greater distance between the public and the increasingly private modes of warfare. War is then no longer a sufficient rallying point for the collectivizing of citizenship. It cannot present a framework for national identity, nor can it serve as a reference point in the quest to understand the pain of others, *even as the violence of drone war in the name of the nation continues*.

In the traditional imagining of the nation-state, the citizen-soldier has come to represent the supreme expression of sacrifice, and therefore citizenship, which other citizens are compelled to aspire to. The figure of the citizen-soldier, making up the first group of people to receive regular benefits, signified ideally proper conduct for all civilians, as people who are seen as sacrificing for a particular body or association are regarded as "true" citizen. However, the paradox of military sacrifice has come to endure a seemingly unbearable pressure. We might trace this back to the end of conscription in some places, coupled with the advent of neoliberal flexible citizenship, or the globalization of military technology, but its most compelling explanatory framework has been the global war on terror – the legal and social innovations on which it is enabled. Rather than liberal democracy's promise of a decline in violence, we have forms of violence with increasingly indefensible and contested meanings. Drone warfare – the technological expression of the sociopolitical philosophy of privatization – is contributing to the expansion of increasingly invisible, and therefore meaningless antisocial violence. Drones and privatization are underpinned by a shared, distancing logic. However, recall the death of Ryan Owens. Of Ryan's death, US president Trump, turning on White House comments regarding the sacrificial nature of his death, remarked that his legacy will be "etched into eternity." Indeed, the power of sacrifice – the desire to make visible certain aspects of war-making – remains deeply entrenched in the narrativization and orientation of liberal politics. Sacrifice is the cornerstone of nation-state violence, and it must find a way to be consolidated in the public imagination. How military deaths are re-presented as those deaths that are etched into eternity tells us something about the past, as well as the future.

If we want drones, however, to tell us something about the future of war, we should look to the past. It might be the case that we no longer need citizen-soldier sacrifice, just like kings and dynasties of prerevolutionary France also did not have much use for citizen-soldiers and instead wielded small and highly trained armies. This was also a kind of antisocial war. Yet we should ask whether this is desirable: Do we want war without citizen-soldiers, without sacrifice, or antisocial war more broadly? For war to be limited and just, liberal democracies must have a clear understanding of citizens, soldiers, and sacrifice and how they are related, particularly in times of war. To mitigate future wars, such an understanding is even more important. To counter the various forms of amnesia baked into privatized drone warfare, we ought to insist that war remain social. To ensure that war remains exceptional, we ought to insist that it maintain its historical memory of its embodied legacy. To ensure war remains political, we must be ever alert to the relationship between distancing technologies and practices, and the nature and quality of wartime wounds, asking to what degree those wounds are publicized or privatized. We also ought to remain ever alert to sacrifice – being perpetually critical and curious about what is at stake when sacrifice is summoned or silenced, demanded or disavowed.

Notes

Chapter 1

1 While an important topic with ongoing implications, this book is not about Ukraine or Russia. Instead, it primarily traces drones in casualty-averse societies, such as the United States. The Russia–Ukraine conflict is clearly a war of attrition, and each side is deploying drones *while also* suffering significant loss of life. In the case of the war on terror, drones emerged in parallel to an ideology of casualty aversion and counterterrorism operations. These are different, albeit related, case studies. However, it is the latter case that is the primary focus of this book.

2 Cindy Sheehan famously became a focal point on the subject of the sacrifice of mothers' of fallen soldiers. While speaking of her own sacrifice as a mother, as well as the sacrifice of her son, she critiqued then president Bush's attempt to minimize her loss. She was able to "mobilize her private suffering in the public domain" (Taussig-Rubbo, 2009, p. 108), which served to destabilize and/or resist a hegemonic interpretation of the public and private spheres, thereby unsettling the gendered expectations of citizenship (Lister, 2002), namely, that she would, as a mother, reserve her grief for the private sphere only and accept the fate of her son as somehow serving a greater, national good. Her claim gained further momentum in her assessment that her son's sacrifice was unnecessary and, therefore, not a legitimate sacrifice, and by virtue of that, meaningless.

3 A historical timeline of sorts is warranted. Veteran's benefits take two forms: commemoration (consider federally owned dedicated cemetery plots, annual days of remembrance, and war medals as examples) and social services. In the United States, benefits gradually moved away from individual state- and community-oriented care. Centralized care under a national framework came about as technology enabled new forms of injury and brutality, and the numbers of veterans escalated. Culminating

in the Servicemen's Readjustment Act of 1944, the GI Bill, (described as one of the most influential pieces of legislation in the United States), was widespread in that it (a) had to account for the significant increase in the veteran population and (b) established medical centres (although it was the First World War – the first mechanized war – that saw the greatest influx of hospitals), made low-interest mortgages available, provided unemployment insurance, and offered stipends for college tuition. Benefits therefore emerged in tandem with the deadliness of contemporary warfare. "Taken together, these well-known facts hint at something unique about the violence intrinsic to modern liberty. This has to do partly with advanced technologies for death dealing. The fact that modern warfare has given birth to numerous inventions is well known. These include improved techniques for destruction, of course, but also for the restoration of human life. Important developments in surgery, psychiatry, and psychology, as well as in nursing and hospital administration, are famously connected with the demands and consequence of modern war" (Asad, 2007, p. 60). It seems possible to invert this relation and question how the increasing "safety" of war for those on "our side" (through enhanced technology and the ideology of casualty aversion) might then also result in the rolling back of veterans' benefits.

4 Isin and Turner (2002) argue that nationality is the primary axis by which people are classified and thus also how citizenship rights can be defined, allocated, and comprehended overall. As such, national trajectories and practices constitute important issues in citizenship studies. The notion of national citizenship, however, has become increasingly problematic because globalization has produced so-called flexible and non-national forms of citizenship and transnational capital flows and supranational organizations have challenged the power of nation-states to effectively govern its citizenry.

5 According to Schuck (2002, p. 132) the meaning and application of liberty varies according to a vast liberal spectrum of ideals. At one extreme, "negative liberty" would advocate the right to total privacy of the individual, whereas "positive liberty" might require the state to actively affirm and create the conditions to secure social entitlements.

6 Isin and Turner (2002), drawing on T.H. Marshall, note that "modern citizenship rights which draw from the nation-state typically include civil (free speech and movement, the rule of law) political (voting, seeking electoral office) and social (welfare, unemployment insurance and health care) rights" (p. 3).

7 Here, it appears that Asad is taking the democratic state's solicitude towards the citizen-soldier as a historical and, perhaps too, a future given. As I attempt to argue throughout this book, there are many examples of

the US state – most notably – retreating from this role as caretaker of the citizen-soldier, through individualizing the citizen-soldier's trauma and emotional suffering and/or withdrawing material benefits, as well as outsourcing military labour to companies whose labourers have no claim to the Department of Veteran's Affairs.

8 Asad (2007) writes, "God's only begotten son gave his life willingly and deliberately in order to redeem mankind: the supreme sacrifice. Although he did not murder himself, he devised that he should be cruelly killed. The Crucifixion has long been a model in Christendom for legal punishment, so that a convicted victim's suffering has been seen as the repayment by which social and metaphysical order can be restored, as a means of cultivating absent virtue, as an example to others of the death that is at once sin and the cleansing of sin" (p. 85).

Chapter 2

1 "Politics begins not in the social contract but in the pledge that expresses a willingness to sacrifice the self for the particular community" (Khan, 2008, p. 111). Khan (2008) goes on to describe this pledge as a "performative utterance" that shows the self-evident truth that is the transcendent value realized in the coming into being of the sovereign community (p. 111).

2 My goal in this regard is not to develop a systematic comparison of Canada and the United States in a straightforward way; rather, it is to highlight points of convergence and divergence. Therefore, I borrow this approach from historical sociologist Philip McMichael (1990). He utilizes comparison in a way in that it concerns the substance of the inquiry rather than the framework. The "whole" that McMichael speaks of emerges via comparative analysis of parts as moments in a self-forming whole. Incorporated comparison allows for the idea that totality is a conceptual procedure, rather than an empirical or conceptual premise. Accordingly, the whole is discovered through an analysis of the mutually conditioning parts. This allows for an analysis of both the material and discursive power dynamics in a way that reveals how the state system is a fluid product of historical processes, which are intimately tied to, and shaped by, social relations.

3 Although there was a moment when the 2003 Iraq war appeared to be just, it was quickly denounced as unjust when it was confirmed that then president Bush's claims about Iraq harbouring weapons of mass destruction were falsified. The continued maintenance of an "enemy" is also critical to making war efforts appear just. As Khan (2011) writes, "even if the ends at stake are just, the absence of an enemy means that there is no legitimate claim on Americans to sacrifice their lives. Nor is there ground for Americans to be killing Iraqis" (p. 156).

4 According to Kahn, sacrifice is loaded with significance. In terms of how it defines the self, "the boundaries of the self only become clear through the imagining of sacrifice, for sacrifice circumscribes the limits of a world … the boundaries of sacrifice are those of love, which is always beyond measure" (p. 109), there is subsequently an implied relation to the other. In this way, sacrifice is also banal (because it is a feature of everyday life). Putting the specifics of this debate aside, Kahn here still does not bring us closer to an understanding of the embodied, uneven, material, relational, and macro-political components of sacrifice that pertain to contemporary war-making practices. Khan is right to point out that states use sacrifice to establish its own borders ("where the state can demand sacrifice is exactly the point at which is establishes its borders"), yet Khan's analysis does not easily lend itself to the changing nature of today's battlefield, the role of technology and privatization, in other words, the differentiated meaning of bodies in warfare today.

5 According to Brighi and Cerella (2015, p. 12) Girard's hypothesis, of "the origin of culture and on the relation between violence and religion, is somewhat controversial because of its *meta-historical* character. The founding murder is in fact an episode (or series of incidents or crises) that happened in *illo tempore*, and cannot be empirically analysed. Yet, according to Girard, there would be *traces*, and significant evidence of this historical moment" (the founding murder).

6 As Feldman (1991) has shown, the deaths of hunger-striking members of the IRA were instances of suicides that were construed as sacrificial (see Taussig-Rubbo, 2009, p. 111) in that they were able to claim sovereignty through the "destruction of a surplus" (p. 118).

7 It is worth citing here the general failure to create a way to honour drone pilots, which would then presumably open the space to in some way memorialize their actions. It is unclear whether Americans regard drone piloting as heroic or not. For example, consider former US defense secretary Leon Panetta's plan in 2013 to make drone pilots eligible for a distinguished warfare medal. As the first combat-related award to be created since the Bronze Star in 1944, it would have ranked higher than both the Bronze Star and Purple Heart, which are typically given to wounded troops. Unlike traditional combat medals, it would not require that the recipient engage in direct combat. Two months following Panetta's announcement, then secretary of state Chuck Hagel, himself a Vietnam veteran with two Purple Hearts, cancelled the medal and replaced it with a distinguishing device that would be affixed to already existing medals. We see here the deepening of a crisis: how to recognize and understand citizen-soldiering today, particularly given the problem postmodern

violence has in reconciling the expanding forms of disembodied violence with liberal democracy's promise of peace.

8 A 2010 report by Mission: Readiness, an organization of retired, senior military experts, warns that childhood and adult obesity might soon pose a threat to national security. It claims that 75 per cent of young Americans are ineligible for military recruitment because of weight, criminal records, asthma, drug abuse, and educational inadequacies.

9 Turning on these developments, and otherwise known as the Bush Doctrine, counterterrorism, as a national strategy, positions the Western state (noble and responsible) as a *target* of violence whose injury resonates at the site of sovereignty in the form of terrorism – illegitimate violence – rather than viewing itself as a practitioner or agent of violence. Its principles, as summarized by Bush himself in his book *Decision Points* (2011) include the following:

"1, Make *no distinction* between terrorists and the nations that harbour them – and hold both to account."

"2. Take the fight to the enemy overseas *before* they can attack us again here at home."

"3. Confront threats *before* they fully materialize."

"4. Advance liberty and hope as an alternative to *the enemy's ideology* of repression and fear."

10 A critical approach to counterterrorism claims that to invoke the word *terrorist* or *terrorism* marks the beginning of an attempt to draw boundaries around various styles and motivations for violence, as well as to provide ground to narrate as well as differentiate violence. The violence of the stateless "other" is considered to be insurgent/barbaric, versus the rationally motivated, sophistically and deliberately calculated, and morally centred violence of the state: the legitimate bearer of violence. Counterterrorism policy, which emerged during the Cold War, homogenizes terroristic violence and categorizes so-called terrorists as unified by a supposedly shared, fanatical hatred for the same things, namely, ideals cherished by the United States and other liberal and/or Western political economies. It historically was just one prong of foreign policy, but now the entirety of military strategy is foreign policy, a conflation that functions to sharply differentiate US-backed terror from the opposing practices it sees as both its justification and inspiration. As the role of the military changed from defence of the homeland to identifying domestic and international security threats and maintaining security both at home and abroad, armed forces in many places are expected to combat threats without enemies, thus broadening the roles of military personnel. Forces are now expected to engage non-traditional military threats: illegal immigration, terrorism, and cyberattack.

Chapter 3

1 Consider Robert N. Bellah's essay on civil religion. He writes that, despite official declarations of secularism, religious imagery, symbolism, and language still circulate in political speeches, particularly in moments of national crises such as in events leading up to war, or those that reflect on war. Sacrifice is an example of a theme that, although profoundly religious in its historical orientation, nevertheless finds a way to be palatable to, and indeed necessary for, both religious and non-religious people alike.

2 As Stuart Hall (2011, p. 708) reminds us, neoliberal ideals come from the principles of 'classic' liberal economic and political theory: over the course of two centuries, "political ideas of 'liberty' became harnessed to economic ideas of the free market: one of liberalism's fault-lines which re-emerges within neoliberalism" (p. 710). Therefore, there is tremendous overlap between so-called liberalism and neoliberalism, both in terms of theory as well as in how they manifest across time and space. Similarly, recall Ong's (1999) definition of liberalism as not "a political philosophy but an art of government ... not something that can be reduced to a perfect realization of a doctrine called liberalism; rather, it includes an array of rationalities whereby a liberal government attempts to resolve problems of how to govern society as a whole" (p. 195). On the subject of how sovereignty operates within each paradigm, consider that they both monumentalize sacrifice, although, within neoliberalism, sacrifice – killing and being killed – is becoming increasingly invisible and thus untethered from its historical emergence in the body of the citizen-soldier, thus the content of sacrifice risks seeming an "animated anachronism."

3 When speaking of liberalism and neoliberalism, one does not involve a complete rejection of the practices of the other. In fact, "neoliberalism ... evolves. It borrows and approximates extensively from classical liberal ideas; but each is given a further 'market' inflexion and conceptual revamp ... neoliberalism performs a massive work of trans-coding while remaining in sight of the lexicon on which it draws" (Hall, 2011, p. 711). Following this passage from Stuart Hall, I use *neoliberalism* to refer to the globalized and marketized *amplification of tensions* contained in classical liberalism. The amplification of these tensions on a global scale is notably reflected in the economic crises characteristic of the post–Cold War period, during which neoliberalism is primarily defined through a language of marketization while not forgetting the "lexicon on which it draws," that is, the spirit of classical liberalism and its emphasis on equality, dignity, and rights for all.

4 Nikolas Rose (1999), in *Powers of Freedom*, utilizes Foucault's notion of governmentality to show how certain positive understandings of freedom

have become a prerequisite for the governing of subjects. To govern is to recognize the capacity for action. Investigations of government focus on "the will to govern as it is enacted in a multitude of programs, strategies, tactics, devices, calculations, negotiations, intrigues, persuasions and seductions aimed at the conduct of the conduct of individuals, groups, populations – and indeed oneself" (p. 5). Furthermore, he claims, "the insistence on the significance of the formation and transformation of truthful thought differentiates studies of government from most varieties of political sociology" (p. 8). For Rose, a genealogy of freedom and the social would examine how relations between freedom and power are established. The point is not to be critical of freedom or reveal it as a grand hoax or think of a more purely truthful freedom but to ask how we have come to define and act towards ourselves in relation to a certain notion of freedom: "The government of freedom can first be analyzed in terms of the inventions of spaces and gazes, the birth of calculated projects, to use space to govern individuals at liberty" (p. 72). Modern individuals are not "free to choose" but "obliged to be free" (p. 87). To efficiently govern is to harness and mobilize those technologies that allow for it to be accomplished at a distance, such that any immediate hierarchy of power is eclipsed.

5 Much like Foucault "denies us a 'theory' of power" (Flynn, quoted in Erlenbusch, 2013, p. 52), it is important to likewise resist a totalizing theory of sacrifice.

6 It is important here to unpack the use of the word *nation* and conceptualize some distinctions between contractual and cultural nations: As ideal types, there is much complicity between the two. As Singer (1996) argues, the cultural nation is thought to be timeless and pre-modern. It uses memory, kinship, language and geography to establish common relations; as such, it can be understood through the terms *particular*, *organic*, and *collective*. In contrast, the contractual nation is directly linked to democracy. It is a political community, defined through the terms *artificial*, *universalist*, and *individualist*.

7 Foucault (2008) writes that the body becomes a useful force as both a productive body and a subjected body. This power projected onto the body is part of a wider political technology of the body, a technology that operates through strategy, which is never a unilateral articulation of power but always diffuse. A strategy is an operation or manifestation of power yet subjects never fully possess power. Foucault imagines the body as a site of relational processes, theorizing materiality as both physical bodies, as well as places, such as the prison. Butler (2004) extends this to say that "it is not just that the movements of the nineteenth century are about the body and material things, as if these two are unrelated objects for such

movements. It is rather that the very materiality of the prison is activated on the body of the prisoner, and through the technology of the soul" (p. 186). Of note is the disjuncture between the institution and the body, and the passage between them. This leads Butler to argue that it is not only the subject by the body itself that is being redefined so that the body is not a substance or a set of drives *but also the site of transfer of power itself.* When Foucault describes the body as material, he means that to be material is not only to resist what works on it but also to be the vector and instrument of a continued working.

8 According to Foucault (2007), since the seventeenth century power has been situated and exercised at the level of life – the life of the social body. This power over life is expressed in Foucault's axis where he posits two poles on a spectrum which were linked together by a variety of relations: "one pole of biopower focuses on an anatomo-politics of the human body, seeking to maximize its forces and integrate it into efficient systems" (p. 139). Moreover, "the second pole is one of regulatory controls, a biopolitics of the population, focusing on the species body, the body imbued with the mechanisms of life: birth, morbidity, morality, longevity" (p. 139). Foucault argues that at the beginning of the nineteenth century, the two poles were conjoined within a series of "great technologies of power," such as sexuality.

9 Michael Hardt and Antonio Negri's (2000, p. 36) concept of Empire refers to a worldwide assemblage wherein some states have significantly more priority than others but one marked above all by the migration of sovereignty towards global structures that exceed the power and control of one single state.

10 Butler (2006, p. 54) states, "petty sovereigns abound, reigning in the midst of bureaucratic army institutions mobilized by aims and tactics of power they do not inaugurate or fully control. And yet such figures are delegated with the power to render unilateral decisions, accountable to no law and without any legitimate authority. The resurrected sovereignty is thus not the sovereignty of unified power under the conditions of legitimacy, the form of power that guarantees the representative status of political institutions."

11 Rosenberg (1994) claims that sovereignty is "highly ambiguous as an actual measure of power" (p. 131).

12 For Sassen (2002), "denationalization" marks the thinning or withdrawal of the state's involvement in citizenship entitlements, as global and supranational governing bodies challenge the efficiency of national economic and political institutions.

13 Benedict Anderson, as discussed by historian Michael J. Allen (2011), argues that nationalism is a secular religion: "For Anderson, nationalism,

like religion, was rooted in the grave. Like all sacred beliefs, nationalism promised eternal life; immortality through survival of the state. One grave in particular stood out as especially meaningful in this regard across a variety of national cultures – the Tomb of the Unknown Soldier. 'No more arresting emblems of the modern culture of nationalism exist than the cenotaphs and tombs of Unknown Soldiers,' he wrote. 'Saturated with ghostly national imaginings,' such shrines were so important to narratives of national immutability that Anderson dared readers to 'imagine the general reaction to the busybody who 'discovered' the Unknown Soldier's name,' calling this seeming impossibility 'sacrilege of a strange and contemporary kind!'" (p. 92).

14 Memorialization practices at the tomb have been highly calculated in the past to the point of fabrication. As Michael J. Allen (2011) claims, allegations were made against the Reagan administration in 1998 that, with respect to the Vietnam Unknown, fourteen years prior, officials had "known the identity of the remains buried there but concealed that information to enact a ritual of national reunion in an election year" (p. 91). The validity of the claims was confirmed in 1998 by *CBS Evening News*, who claimed, "documents dug out by CBS clearly indicate that Reagan officials and the US military knew the remains … when they were buried … and may have covered that up" (p. 91).

15 In his landmark essay, Bellah (1967) considers how a profound heritage of moral and religious experiences in America can be redistributed and re-presented publicly in a formally secular era. Of this secular era, Angrosino (2002) writes, "social scientists, heir to the positivist traditions of Comte and Marx, accepted as given the trend of modern societies towards 'secularization,' and hence have grown increasingly impatient with the notion that religion- even a 'civil' one- has any place in a modern polity" (p. 240). Furthermore, as Hammond, quoted in Angrosino, writes, in this context, "the public square does not rule out religious rules and motives; it simply does not accord them authority until they are translated [into terms readily understandable even by the non-religious]" (p. 242). Bellah defined civil religion as "a collection of beliefs, symbols, and rituals with respect to sacred things and institutionalized in a collectivity. This religion – there seems to be no other word for it – while not antithetical to and, indeed, sharing much in common with Christianity, was neither sectarian nor in any specific sense Christian … it reflected their [society's] private as well as public views. Nor was the civil religion simply 'religion in general …' the religion was specific enough when it came to the topic of America" (p. 8). Angrosino adds that "American civil religion is an institutionalized set of beliefs about the nation, including a faith in a transcendent deity who will protect and guide the United States as long

as its people and government abide by its laws" (p. 241). Bellah asks how we are supposed to determine the references to God in political speeches, when we find "references to God are almost invariably to be found in the pronouncements of American presidents on solemn occasions, though usually not in the working messages that the President sends to Congress on various concrete issues" (p. 2). In countering the idea that God has only a "ceremonial significance" and/or the cynical viewpoint that reference to God is used purely for political gain, Bellah demonstrates how the separation of church and state has not denied the public sphere a religious dimension (this public dimension is the very notion of American civil religion).

Chapter 4

1 After fifteen months of arbitration, the Fifth Circuit Court of Appeals ruled Jones's federal lawsuit against KBR and several affiliates could be tried in open court. Jones was finally able to sue KBR in court, thereby effectively moving her claims of suffering from the private sphere of the boardroom into the public domain of the trial. She filed a civil suit against KBR and requested US$114 million dollars in damages. On 8 July 2011, the federal jury ruled in favour of KBR. They found that the sex with the accused, firefighter Charles Bortz, was consensual and thus no rape occurred. The jury also concluded that KBR did not defraud Jones.

2 Consider the war in Bosnia in 1999. DynCorp employees, who made up the core of the police force, were reported to have raped young girls, videotaping their activities. Moreover, they allegedly participated in sex-trafficking schemes, including buying and selling women for their own personal enjoyment, in addition to purchasing illegal weapons, forged passports and committing other illegal acts (Bolkovac & Lynn, 2011).

3 The Bush administration never publicly articulated itself as feminist in orientation prior to 9/11, but it co-opted wholesome-sounding liberal feminist rhetoric to justify its imperialist agenda via "embedded feminism" (Rygiel, 2008b). Feminists remain interested in why, in the historical juncture of the post-9/11 period, women's rights discourses were popularized by a neoconservative administration that had otherwise sought to aggressively dismantle the gains that feminists made in the past thirty years.

4 The use of "republican" here has a distinct meaning (when compared to how the term is deployed inside the United States) as it specifically applies to the history of the Irish "Troubles." At the heart of the thirty-year conflict (1968–1988) lay the constitutional status of Northern Ireland. While the unionists (Protestant majority) sought to remain part of the United

Kingdom, the nationalist, republican, and almost exclusively Catholic minority was to become part of the Republic of Ireland.

5 Although compelling, this viewpoint nonetheless requires clarification, particularly because it directly conflicts with official United States policy, which indicates it prefers capture operations to lethal action. In the "Drone Memos," dated 22 May 2013, standard operating procedure states for targeting terrorists outside of the United States that "capture operations offer the best opportunity for meaningful intelligence gain from counterterrorism (CT) operations and the mitigation and disruption of terrorist threats. Consequently, the United States prioritizes, as a matter of policy, the capture of terrorist suspects as a preferred option over lethal action and it will therefore require a feasibility requirement of capture options as a component of any proposal for lethal action. Lethal action should be taken to prevent terrorist attacks against US persons only when the capture of an individual is not feasible and no other reasonable alternatives exist to effectively address the threat. Lethal action should not be proposed or pursued as a punitive step or as a substitute for prosecuting a terrorist suspect in a civilian court or a military commission. Capture is preferred even in circumstances where neither prosecution or third-country custody are available disposition options at the time." In some instances, while capture might be preferred, the suspect ends up being killed. Such was the case in March 2016 when Special Forces killed the Islamic State of Iraq and the Levant's "oil minister" Abu Sayyaf. They had intended to take the leader alive, but he was killed when he attempted to fight back (Ensor, 2017).

Chapter 5

1 The invention and application of the A-4/V-2 rocket during the Second World War, the first long-range ballistic missile to be used in combat, provides a counter-narrative of the one offered by this interviewee.

2 "Due to limited fuel capacity, jets normally can't observe a target for too long before a strike. However, drones can stay over a target for a long time. This means that the drone can be used to observe a target and wait for a clear shot in situations where civilian losses are being minimalized. Drones can also be used in signature strikes, meaning that instead of targeting a specific person, a drone will fly over an area and look for signs of militant activity. If a certain amount of militant signatures are observed, the drone can fire even if the operator doesn't know the exact identity of the target(s)" (AJ, Interview).

3 "From the time you go into the military you're always trained, and it's really driven into that, you go where you told, you don't get to choose,

you're there to serve and that's just the way it is: 'mission focus' is a phrase that permeates" (AS, Interview).

4 This desire to avoid being surprised (by doing the surprising) is confirmed in a separate interview data by a DARPA employee: "DARPA's role is to look beyond immediate national security needs and enable future technological surprise by investing in breakthrough technologies." (The views and opinions expressed should not be interpreted as expressing the official views or policies, either expressed or implied, of the Defense Advanced Research Projects Agency or the Department of Defense).

5 This interlocutor's confused use of geography is not lost on the author. I believe the main point stands.

6 It is not simply the Vietnam War that characterizes the link between casualty avoidance and high technology. It is the loss of life during wartime in general, specifically as read through air power. As TW explains, "If you go to Edwards Air Force Base all the roads are named after guys who got killed in flight tests; there haven't been too many added since the '60s, there have been a few, that's in proportion or percentage wise that were killed in flight tests in the '80s and '90s but not anywhere close to the number of folks that were killed in the early days of supersonic testing, and the same thing in the fighter world, airplanes have gotten a lot more sophisticated, weapons have gotten a lot more sophisticated, and the night capability has increased significantly and weapons' precision is just phenomenal" (Interview).

7 "It's really about the transformation of military logistics, over the last few decades, particularly post-Vietnam, the US government has been trying to figure out how to, for reasons of cost, but also for reasons of morale, to shrink the size of the US army. Kids don't want to go to war. They don't understand the need for it, so when you send people to war using a draft you're gonna get some very reluctant people who are maybe unruly; a lot of people deserted, a lot of people messed up with drugs, a lot of people even shot their officers, so you had a real problem in trying to get people to the war" (PC, Interview).

8 Note that in interview excerpts, emphasis has been added to highlight important points.

9 "Your average combat air patrol, which involves three to four drones, has 186 people involved in it. Now that is the predator or reaper, if it's a global hawk, it's a lot more people. And those range, from the launch and recovery element, which is the people that actually put it up in the air, and they use what's called line of sight technology to get it going and bring it back, and then once it's in the air you have a pilot who is typically elsewhere, not necessarily at the launch site. If I can narrow this let's say

to a predator in use over Pakistan let's say, that is typically managed not by the launch and recovery pilot who is in let's say Afghanistan, but it's managed by somebody, let's say, in Nevada, now, they obviously cannot launch it, that is done by somebody at the physical base, but they manage it once it goes beyond line of sight, essentially managed through what's called BLS (beyond line of sight). So then you have a pilot, and you have a camera operator. So in theory there are two people that physically operate it: the person who manages the cameras and the person who manages the plane, however in fact there's a lot more people involved" (PC, Interview).

10 "Not a lot of people want to join this [drone operations] and there's a variety of reasons, because it's a catch-22, they have so few, and the military they're 65 combat air patrols (CPS) and they want to increase it to 90. It takes 12–18 months to get somebody ready to do these tasks they can only train so many, so they are shorthanded, these people are working 12 hours a day, seven days a week. So most people when they can get out they get out simply because they are exhausted, so the reason why this is very difficult is because they're so short staffed, so they're pulling other people in from other services, air frames, etc. so there's been a huge amount of attrition purely because they don't have enough so let's suppose in a perfect world they have the perfect number of pilots that problem would go away, what they've been doing is trying to increase pilots, privatize it to a certain extent, bring back people who have retired, provide counselling to people who have had problems with the distance between their two lives etc., and that is essentially solvable, it's somewhat ironic because there are these whistleblowers, perhaps if they had a cushy life and didn't have to work that much they wouldn't complain at all" (PC, Interview).

11 "Imagery analysts who review video footage are among the lowest ranked among the personnel who work in the drone war hierarchy. Typically these are entry level 'airmen' who only need a high school diploma and eleven months of military training. The drone pilots are officers with undergraduate degrees and more years of training" (Chatterjee, 2015).

12 "A spokeswoman for the Air Force said intelligence, surveillance, and reconnaissance (ISR) was 'vital to the national security of the United States and its allies,' and there was an 'insatiable demand' for it from combatant commanders. She said this demand was the reason for increasing use of contractors, which she said was a 'normal process within military operations'" (Fielding-Smith & Black, 2015).

13 For example, the task of simply knowing how many people are involved in private contracts remains impossible. A report titled *Contractors' Support of Operations in Iraq*, produced by the US Congressional Budget Office (2008), found that, "increased use of performance-based contracting – a

procurement model that emphasizes outcomes rather than specification of work processes – has further reduced the government's ability to know how many people are working on a given contract."

14 "The Air Force Distributed Common Ground System (AF DCGS) is the Air Force's primary intelligence, surveillance and reconnaissance (ISR) planning and direction, collection, processing and exploitation, analysis and dissemination (PCPAD) weapon system. The weapon system employs a global communications architecture that connects multiple intelligence platforms and sensors. Airmen assigned to AF DCGS produce actionable intelligence from data collected by a variety of sensors on the U-2, RQ-4 Global Hawk, MQ-1 Predator, MQ-9 Reaper and other ISR platforms" (US Air Force, n.d.).

15 Meanwhile, the USAF is currently tackling the problematic link between stressful working conditions and its impacts on job efficiency and specifically, mechanical memory. Historically, it has used Ritalin to treat a variety of mental health disorders in the services but has found it to be as addictive as cocaine and has since attempted to develop alternatives. As a result, it is currently experimenting with "transcranial direct current stimulation." Neuroscientists and military researchers have developed a brain stimulation kit consisting of five electrodes that send weak electric currents into specific parts of the cerebral cortex. "The US military has found over the past few decades that during highly stressful military operations whereby people need to multitask and process a huge amount of information at once, there comes a point whereby human operators of specialised equipment can no longer take on any new information and process it effectively to make good decisions quickly" (Russon, 2016). From their experiments, they concluded that the group that was continuously zapped had improved working memory.

References

Abdullah, T. (2003). *A short history of Iraq: From 636 to the present*. London: Pearson/Longman. https://doi.org/10.4324/9781315834214

Ackerman, S., Burke, J., & MacAskill, E. (2017, February 2). Questions mount over botched Yemen raid approved by Trump. *The Guardian*. Retrieved from www.theguardian.com/world/2017/feb/02/trump-approved-yemen-raid-five-days-after-inauguration

Adamsky, D. (2010). *The culture of military innovation*. Stanford, CA: Stanford University Press. https://doi.org/10.1515/9780804773805

Adey, P., Whitehead, M., & Williams, A. J. (2013). *From above: War, violence and verticality*. Oxford: Oxford University Press. https://doi.org/10.1093/acprof:oso/9780199334797.001.0001

Agamben, G. (1998). *Homo Sacer: Sovereign power and bare life* (D. Heller-Roazen, Trans.). Stanford, CA: Stanford University Press. https://doi.org/10.1515/9780804764025

Agamben, G. (2002). *Remnants of Auschwitz* (D. Heller-Roazen, Trans.). New York, NY: Zone Books. https://doi.org/10.1111/j.1467-8330.2011.00940.x

Akhter, M., & Shaw, I. G. R. (2012). The unbearable humanness of drone warfare in FATA, Pakistan. *Antipode, 44*(4), 1490–1509. https://doi.org/10.1111/j.1467-8330.2011.00940.x

Alexandra, A., Baker, D., & Caparini, M. (Eds.). (2008). *Private military and security companies: Ethics, policies and civil-military relations*. New York, NY: Routledge. https://doi.org/10.4324/9780203930830

Allen, M. J. (2011). 'Sacrilege of a strange and contemporary kind': The Unknown Soldier and the imagined community at Vietnam. *History and Memory, 23*(2), 90–131. https://doi.org/10.2979/histmemo.23.2.90

Al Sane, N., & Shabibi, N. (2017, September 2). Nine young children killed: The full details of botched US raid in Yemen. *The Bureau of Investigative Journalism*. Retrieved from www.thebureauinvestigates.com/stories/2017-02-09/nine-young-children-killed-the-full-details-of-botched-us-raid-in-yemen

Amoore, L., (2020). *Cloud ethics: Algorithms and the attributes of ourselves and others.* Durham, NC: Duke University Press. https://doi.org/10.1215/9781478009276

Anderson, B. R. O. (1991). *Imagined communities: Reflections on the origin and spread of nationalism* (Rev. ed.). New York, NY: Verso.

Angrosino, M. (2002). Civil religion redux. *Anthropological Quarterly, 75*(2), 239–267. http://www.jstor.org/stable/3318260

Appadurai, A. (1996). *Modernity at large: Cultural dimensions of globalization.* Minnesota, MN: University of Minnesota Press.

Aradau, C., & Blanke, T. (2022). *Algorithmic reason: The new government of self and other.* Oxford, UK: Oxford University Press. https://doi.org/10.1093/oso/9780192859624.003.0001

Arkin, R. C. (2009). *Governing lethal behavior in autonomous robots.* Boca Raton, FL: CRC Press. https://doi.org/10.1201/9781420085952

Arrighi, G. (2007). *Adam Smith in Beijing.* New York, NY: Verso. https://doi.org/10.1017/S1598240800006986

Asad, T. (2007). *On suicide bombing.* New York, NY: Columbia University Press.

Asaro, P. M. (2013). The labor of surveillance and bureaucratized killing: New subjectivities of military drone operators. *Social Semiotics, 23*(2), 196–224. https://doi.org/10.1080/10350330.2013.777591

Atkins, E., & Seamone, E. R. (2024). Remote combat exposure and moral injury from drone operations: The cost of a new form of warfare. In J. T. McDaniel, E. R. Seamone, & S. N Xenakis (Eds)., *Preventing and treating the invisible wounds of war: Combat trauma, moral injury, and psychological health* (pp. 207–236). Oxford, UK: Oxford Academic. https://doi.org/10/1093/oso/9780197646588.003.0010

Austen, I. (2012, October 7). Canada puts spotlight on war of 1812, with U.S. as villain. *The New York Times.* https://mobile.nytimes.com/2012/10/08/world/americas/canada-highlights-war-of-1812-casting-us-as-aggressor.html

Australian Government, Defence. (2024, February 9). *Albanese government invest further $400 million in next generation loyal wingman drone* [Press release]. https://www.minister.defence.gov.au/media-releases/2024-02-09/albanese-government-invests-further-400-million-next-generation-loyal-wingman-drone

Avant, D. (2007). The emerging market for private military services and the problems of regulation. In S. Chesterman & C. Lehnardt (Eds.), *From mercenaries to market* (pp. 181–195). Oxford: Oxford University Press.

Avant, D., & Sigelman, L. (2010). Private security and democracy: Lessons from the US in Iraq. *Security Studies, 19*(2), 230–265. https://doi.org/10.1080/09636412.2010.480906

Avant, D. D. (2005). *The market for force: The consequences of privatizing security.* New York, NY: Cambridge University Press.

Baggiarini, B. (2015). Drone warfare and the limits of sacrifice. *Journal of International Political Theory, 11*(1), 128–144. https://doi.org/10.1177/1755088214555597

Baggiarini, B. (2019). Governing the social through sacrifice: Militarization, commemoration, and Canadian identity. In D. Brock (Ed.), *Governing the social in neoliberal times* (pp. 242–260). Vancouver: University of British Columbia Press. https://doi.org/10.59962/9780774860925-012

Baggiarini, B. (2024). Algorithmic war and the dangers of in-visibility, anonymity, and fragmentation. *Australian Journal of International Affairs, 78*(2), 257–265. https://doi.org/10.1080/10357718.2024.2333824

Baggiarini, B., & Rupka, S. (2020). Remembering the drone wars: high technology warfare and the trauma lacuna. *Social Research: An International Quarterly, 87*(3), 769–786. https://dx.doi.org/10.1353/sor.2020.0061

Balakrishnan, G. (2009). *Antagonistics: Capitalism and power in an age of war.* London: Verso.

Balibar, E. (2011). The nation form: History and ideology. In E. Balibar & I. Wallerstein (Eds.), *Race, nation, class* (pp. 86–106). New York, NY: Verso.

Barker, I. V. (2009). (Re)producing American soldiers in an age of empire. *Politics & Gender, 5*(2), 211–235. https://doi.org/10.1017/S1743923X090000166

Barker, I. (2015). (Re)producing American soldiers in an age of empire. In M. Eichler (Ed.), *Gender and private security in world politics* (pp. 75–95). Oxford: Oxford University Press. https://doi.org/10.1093/acprof:oso/9780199364374.001.0001

Baker, P., Hirschfield Davis, J., &Shear, M. (2017, February 28). Trump, in optimistic address, asks congress to end 'trivial fights.' *New York Times*. Retrieved from www.nytimes.com/2017/02/28/us/politics/trump-address-congress.html

Bataille, G., & Hurley, R. (1989). *The accursed share, volume I.* New York: Zone Books. https://doi.org/10.2307/jj.20123091

BBC. (2023, 29 December). *How are 'kamikaze' drones being used by Russia and Ukraine?* https://www.bbc.com/news/world-62225830

Beck, U. (2000). *What is globalization*? (P. Camiller, Trans.). Malden, MA: Polity Press. (Original work published 1997)

Beier, J. M. (2012). Dangerous terrain: Re-reading the landmines ban through the social worlds of the RMA. In N. Cooper & D. Mutimer (Eds.), *Reconceptualising arms control: Controlling the means of violence* (pp. 157–173). New York, NY: Routledge.

Bellah, R. N. (1967). Civil religion in America. *Dædalus, 96*(1), 1–21. http: www.jstor.org/stable/20027022

Benhabib, S. (2004). *The rights of others: Aliens, residents and citizens.* Cambridge, MA: Cambridge University Press. https://doi.org/10.1017/CBO9780511790799

Benjamin, M. (2012). *Drone warfare: Killing by remote control.* New York, NY: OR Books.

Blanchard, E. (2011). The technoscience question in feminist international relations: Unmanning the US war on terror. In J. Ann Tickner & L. Sjoberg (Eds.), *Feminism and international relations: Conversations about the past, present, and future* (pp. 146–168). London and New York, NY: Routledge. https://doi.org/10.4324/9780203816813

Block, F. (2001). Introduction. In K. Polanyi (Ed.), *The great transformation: The political and economic origins of our time* (pp. xviii–xxxvi). Boston, MA: Beacon Press.

Bolkovac, K., & Lynn, C. (2011). *The whistleblower: Sex trafficking, military contractors, and one woman's fight for justice.* New York: St. Martin's Press.

Bousquet, A. (2018). *The eye of war: Military perception from the telescope to the drone.* University of Minnesota Press. https://doi.org/10.5749/j.ctv6hp332

Brodie, J. (2008). The social in social citizenship. In E. Isin (Ed.), *Recasting the social in citizenship* (pp. 20–44). Toronto: University of Toronto Press. https://doi.org/10.3138/9781442688957

Brown, J. K. (2017, March 1). Slain SEAL's dad wants answers: 'Don't hide behind my son's death. *Miami Herald.* https://www.miamiherald.com/news/politics-government/article135064074.html

Brown, W. (1995). *States of injury: Power and freedom in late modernity.* Princeton, NJ: Princeton University Press. https://doi.org/10.1515/9780691201399

Broze, D. (2016, July 7). Is the CIA manipulating the weather? *Activist Post.* Retrieved from www.activistpost.com/2016/07/is-the-cia-manipulating-the-weather.html.

Brubaker, R. (1992). *Citizenship and nationhood in France and Germany.* Cambridge, MA: Harvard University Press. https://doi.org/10.4159/9780674028944

Burchell, D. (2002). Ancient citizenship and its inheritors. In E. Isin & B. Turner (Eds.), *Handbook of citizenship studies* (pp. 84–104). London: Sage. https://doi.org/10.4135/9781848608276

Burchell, G., Gordon, C., & Miller, P. (Eds.). *The Foucault effect: Studies in governmentality.* Chicago: University of Chicago Press. https://press.uchicago.edu/ucp/books/book/chicago/F/bo3684463.html

Butler, J. (2004). *Precarious life: The powers of mourning and violence.* London: Verso.

Canadian Heritage. (2014). *Commemorating and honouring a legacy of service and sacrifice* [News release]. https://www.canada.ca/en/news/archive/2014/01/commemorating-honouring-legacy-service-sacrifice.html

Cavallaro, J., Knuckey, S., & Sonnenberg, S. (2012, September 25). *Living under drones: Death, injury, and trauma to civilians from U.S. drone practices in Pakistan.* Stanford, NY: International Human Rights and Conflict Resolution Clinic, Stanford Law School; NYS School of Law, Global Justice Clinic. Retrieved from https://law.stanford.edu/publications/living-under-drones-death-injury-and-trauma-to-civilians-from-us-drone-practices-in-pakistan/

Cerella, A., & Brighi, E. (2015). An alternative vision of politics and violence: Introducing mimetic theory in international studies. *Journal of International Political Theory, 11*(1), 3–25. https://doi.org/10.1177/1755088214555455

Chamayou, G. (2015). *Drone theory.* UK: Penguin Books.

Chatterjee, P. (2004). *Iraq Inc.: A profitable occupation.* New York, NY: Seven Stories Press.

Chatterjee, P. (2015, 3 August). *Outsourcing the kill chain: Eleven drone contractors revealed.* CorpWatch. https://www.corpwatch.org/article/outsourcing-kill-chain-eleven-drone-contractors-revealed

Ciccarelli, R. (2008). Reframing political freedom: An analysis of governmentality. *European Journal of Legal Studies, 1*(3), 1–27.

Clark, J. (2023, August 28). *Hicks underscores U.S. innovation in unveiling strategy to counter China's military buildup.* U.S. Department of Defense. https://www.defense.gov/News/News-Stories/Article/Article/3507514/hicks-underscores-us-innovation-in-unveiling-strategy-to-counter-chinas-militar/

Clark, L. (2019). *Gender and drone warfare: A hauntological perspective* (1st ed.). Routledge. https://doi.org/10.4324/9780429507472

Clarke, T. (1996). Mechanisms of corporate rule. In J. Mander & E. Goldsmith (Eds.), *The case against the global economy: And for a turn toward the local* (pp. 297–308). San Francisco, CA: Sierra Club Books. https://doi.org/10.4324/9781315071787

Clausewitz, C. v. (1976). *On war* (M. Howard & P. Paret, Eds.). Princeton, NJ: Princeton University Press.

Cloud, D. S. (2011, April 10). Anatomy of an Afghan war tragedy. *LA Times.* Retrieved from http://articles.latimes.com/2011/apr/10/world/la-fg-afghanistan-drone-20110410

Cloud, D. S., & Zucchino, D. (2011, October 14). U.S. deaths in drone strikes due to miscommunication, report says. *LA Times.* Retrieved from http://articles.latimes.com/2011/oct/14/world/la-fg-pentagon-drone-20111014

Coeckelbergh, M. (2013). Drones, information technology, and distance: Mapping the moral epistemology of remote fighting. *Ethics and Information Technology, 15*(2), 87–98. Retrieved from https://link.springer.com/article/10.1007/s10676-013-9313-6

Cohn, C. (1987). Sex and death in the rational world of defense intellectuals. *Signs: Journal of Women in Culture and Society, 12*(4), 687–718. https://www.journals.uchicago.edu/doi/abs/10.1086/494362

Coll, S. (2014, November 24). The unblinking stare. *The New Yorker.* Retrieved from www.newyorker.com/magazine/2014/11/24/unblinking-stare

Connolly, W. E. (2004). The complexity of sovereignty. In J. Edkins, V. Pin-Fat, & J. M. Shapiro (Eds.), *Sovereign lives: Power in global politics* (pp. 23–40). New York, NY: Routledge. https://doi.org/10.4324/9780203338735

Cooper, R. (2021, 1 December). Biden nearly ended the drone war, and nobody noticed. *The Week.* https://theweek.com/foreign-policy/1007579/biden-nearly-ended-the-drone-war-and-nobody-noticed

Cowen, D. (2008). *Military workfare: The soldier and social citizenship in Canada.* Toronto, ON: University of Toronto Press.

Cowen, D., & Gilbert, E. (Eds.). (2008). *War, citizenship, territory.* New York, NY: Routledge. https://doi.org/10.1080/00045601003638808

Crawford, K. (2021). *The atlas of AI: Power, politics, and the planetary costs of artificial intelligence.* New Haven, CT: Yale University Press. https://doi.org/10.1080/23738871.2023.2237981

Dagger, R. (2002). Republican citizenship. In E. Isin & B. Turner (Eds.), *Handbook of citizenship studies.* London: Sage. https://doi.org/10.4135/9781848608276

Dalby, S. (2008). *Geopolitics, the revolution in military affairs, and the Bush doctrine.* YCISS Working Paper (49). York Centre for International Security Studies, Toronto, ON.

Dao, J. (2013, February 22). Drone pilots are found to get stress disorders much as those in combat do. *The New York Times.* Retrieved from www.nytimes.com/2013/02/23/us/drone-pilots-found-to-get-stress-disorders-much-as-those-in-combat-do.html

Dauphinée, E., & Masters, C. (Eds.). (2007). *The logics of biopower and the war on terror: Living, dying, surviving.* New York, NY: Palgrave Macmillan. https://doi.org/10.1007/978-1-137-04379-5

Davis, M. (2006). *Planet of slums.* London: Verso.

Dean, M. (1999). *Governmentality: Power and rule in modern society.* Thousand Oaks, CA: Sage.

Decker, A. (2023, March 7). *Air Force wants more planes—Plus a thousand drone wingmen.* Defense One. https://www.defenseone.com/policy/2023/03/air-force-wants-more-planes-fasterplus-thousand-drone-wingmen/383687/

De Landa, M. (1991). *War in the age of intelligence.* New York: Zone Books.

Della Volpe, J. (2015, December 10). Survey of young Americans' attitudes toward politics and public service. *Harvard University Institute of Politics.* Retrieved from www.iop.harvard.edu/survey-young-americans%E2%80%99-attitude-toward-politics-and-public-service-24th-edition

Delphy, C. (2003). A war for Afghan women? In S. Hawthorne & B. Winter (Eds.), *After shock: September 11, 2001, global feminist perspectives* (pp. 27–42). Vancouver, BC: Raincoast Books.

Der Derian, J. (2000). Virtuous war/virtual theory. *International Affairs, 76*(4), 771–788. http://www.jstor.org/stable/2626459

Der Derian, J. (2001). *Virtuous war.* Colorado: Westview Press.

Dickinson, L. (2015, July 30). Reaping the rewards: How the private sector is cashing in on the Pentagon's 'insatiable demand' for drone war intelligence. *Bureau of Investigative Journalism.* Retrieved from www.thebureauinvestigates.com/stories/2015-07-30/reaping-the-rewards-how-private-sector-is-cashing-in-on-pentagons-insatiable-demand-for-drone-war-intelligence

Drew, C., & Philipps, D. (2015, June 16). As stress drives off drone operators, Air Force must cut flights. *The New York Times*. Retrieved from www.nytimes.com/2015/06/17/us/as-stress-drives-off-drone-operators-air-force-must-cut-flights.html

Dudziak, M. L. (2012). *War time: An idea, its history, its consequences*. New York: Oxford University Press.

Duffield, M. R. (2001). *Global governance and the new wars: The merging of development and security*. London: Zed Books.

Du Gay, P., & Hall, S. (Eds.). (2011). *Questions of cultural identity*. London: Sage. https://doi.org/10.4135/9781446221907

Edkins, J. (2003) *Trauma and the memory of politics*. Cambridge, UK: Cambridge University Press. https://doi.org/10.1017/CBO9780511840470

Edkins, J., Pin-Fat, V., & Shapiro, M. J. (2004). *Sovereign lives: Power in global politics*. New York, NY: Routledge. https://doi.org/10.4324/9780203338735

Eichler, M. (2015). *Gender and private security in global politics*. Oxford: Oxford University Press. https://doi.org/10.1080/17502977.2015.1134400

Elshtain, J. B. (1993). Sovereignty, identity, sacrifice. In M. Ringrose, & A. J. Lerner (Eds.), *Reimagining the nation* (pp. 159–175). Philadelphia, PA: Open Press. https://doi.org/10.1177/030582989102000309O1

Elshtain, J. B. (2019). The compassionate warrior: Wartime sacrifice. In R. J. González, H. Gusterson, & G. Houtman (Eds.), *Militarization: A reader* (pp. 91–94). Durham, NC: Duke University Press. https://doi.org/10.1215/9781478007135

Enemark, C. (2019). Drones, risk, and moral injury. *Critical Military Studies*, *5*(2), 150–167. https://doi.org/10.1080/23337486.2017.1384979

Enemark, C. (Ed.). (2021). *Ethics of drone strikes: Restraining remote-control killing*. Edinburgh University Press. https://www.jstor.org/stable/10.3366/j.ctv1c29rjh

Enemark, C. (2023). *Moralities of drone violence*. Edinburgh University Press. https://library.oapen.org/handle/20.500.12657/62317

Enloe, C. (2019). Wounds: Militarized nursing, feminist curiosity, and unending war. *International Relations*, *33*(3), 393–412. https://doi.org/10.1177/0047117819865999

Ensor, J. (2016, January 9). U.S. special forces carry out secret ground raid against Isil in Syria, 'killing at least 25 jihadists'. *The Telegraph*. Retrieved from www.telegraph.co.uk/news/2017/01/09/us-special-forces-carry-ground-raid-against-isil/

Erlenbusch, V. (2013). The place of sovereignty: Mapping power with Agamben, Butler, and Foucault. *Critical Horizons*, *14*, 44–69. https://doi.org/10.1179/15685160X13A.0000000003

Fairclough, N. (1989). *Language and power*. London: Longman.

Feldman, A. (1991). *Formations of violence*. Chicago, IL: University of Chicago Press.

Fielding-Smith, A., & Black, C. (2015, July 30). *Reaping the rewards: How private sector is cashing in on Pentagon's 'insatiable demand' for drone war intelligence.* The Bureau of Investigative Journalism. https://www.thebureauinvestigates.com/stories/2015-07-30/reaping-the-rewards-how-private-sector-is-cashing-in-on-pentagons-insatiable-demand-for-drone-war-intelligence

Fielding-Smith, A., & Serle, J. (2014, October 3). 'It's very odd:' A former U.K drone operator speaks. *Bureau of Investigative Journalism*. Retrieved from www.thebureauinvestigates.com/stories/2014-10-03/its-very-odd-a-former-uk-drone-operator-speaksFoucault, M. (1978). *History of sexuality, Vol.1: An introduction* (R. Hurley, Trans.) New York, NY: Random House.

Foucault, M. (1991). Governmentality. In G. Burchell, C. Gordon, & P. Miller (Eds.), *The Foucault Effect: Studies in governmentality* (pp. 87–104). Chicago, IL: University of Chicago Press.

Foucault, M. (2003). *Society must be defended: Lectures at the Collège de France, (1975–1976)* (M. Beraini & A. Fontana, Eds.). New York, NY: Picador.

Foucault, M. (2007). *Security, territory, population: Lectures at the Collège de France, (1977–1978)* (M. Senellart, Ed., G. Burchell, Trans.). Basingstoke: Palgrave Macmillan.

Foucault, M. (2008). *The birth of biopolitics: Lectures at the Collège de France, (1978–1979)* (M. Senellart, Ed., G. Burchell, Trans.). Basingstoke: Palgrave Macmillan.

Fournier, P. (2014, May 12). *Foucault and international relations*. E-International Relations. https://www.e-ir.info/2014/05/12/foucault-and-international-relations/

Franke, U. (2025, January 10). *Drone in Ukraine: Four lessons for the West*. European Council on Foreign Relations. https://ecfr.eu/article/drones-in-ukraine-four-lessons-for-the-west/

Freeman, J., & Minow, M. (2009). *Government by contract: Outsourcing and American democracy*. Cambridge, MA: Harvard University Press. https://doi.org/10.2307/j.ctv22jnrsb

Garamone, J. (2023, September 7). *Hicks discusses replicator initiative*. U.S. Department of Defense. https://www.defense.gov/News/News-Stories/Article/article/3518827/hicks-discusses-replicator-initiative/

George, L. (2002). The pharmacotic war on terrorism: Cure or poison for the US body politic? *Theory, Culture, and Society, 19*(4), 161–186. https://doi.org/10.1177/0263276402019004012

Gill, S. (2005). The contradtictions of U.S. supremacy. In L. Panitch & C. Leys (Eds.), *Empire reloaded: Socialist register 2005* (pp. 34–56). Halifax, NS: Fernwood Publishing.

Girard, R. (1979). *Violence and the sacred* (P. Gregory, Trans.) Baltimore, MD: Hopkins Fulfillment Service.

Government of Canada. Office of the Minister of Canadian Heritage. (2011). *Harper government launches the commemoration of the 200th anniversary of the war of 1812* [Press release]. Retrieved from www.canada.ca/en/news/archive/2011/10/harper-government-launches-commemoration-200th-anniversary-war-1812.html

Graham, S. (2008). Imagining urban warfare. In D. Cowen & E. Gilbert (Eds.), *War, citizenship, territory* (pp. 33–57). New York, NY: Routledge.

Gramsci, A. (1971). *Selections from "The Prison Notebooks."* (Q. Hoare & G. N. Smith, Trans.). Newark, NJ. International Publishers. (Original works written 1932–1939 and published circa 1950).

Gray, C. (2003). Posthuman soldiers in postmodern war. *Body and Society, 9*(4), 215–226. https://doi.org/10.1177/1357034O3773684739

Grayson, K. (2016). *The cultural politics of targeted killing On drones, counter-insurgency, and violence.* Abingdon, UK: Taylor & Francis.

Green, A. (2024, February 8). *Funding boost for lethal Ghost Bat drone project as Labor confirms armed drone will be introduced this year.* ABC News (Australia). https://www.abc.net.au/news/2024-02-09/funding-boost-for-lethal-ghost-bat-drone-project/103442292

Greenwald, G. (2013, May 1). A young Yemini writer on the impact and morality of drone-bombing his country. *The Guardian*. Retrieved from www.theguardian.com/commentisfree/2013/may/01/ibrahim-mothana-yemen-drones-obama

Gregory, D. (2011). The everywhere war. *The Geographic Journal, 177*(3), 238–250. http://www.jstore.org/stable/41238044

Gregory, D. (2017). DIRTY DANCING. Drones and death in the borderlands. In L. Parks & C. Kaplan (Eds.), *Life in the Age of Drone Warfare* (pp. 25–58). Duke University Press. https://doi.org/10.2307/j.ctv121033r.5.

Gusterson, H. (2016). *Drone: Remote control warfare.* The MIT Press. https://www.jstor.org/stable/j.ctt1c3gwxg

Hall, R. (2007). Of Ziploc bags and black holes: the aesthetics of transparency in the war on terror. *The Communication Review, 10*(4), 319–346. https://doi.org/10.1080/10714420701715381

Hardt, M., & Negri, A. (2000). *Empire.* Cambridge, MA: Harvard University Press.

Harper, S. (2014, 4 August). *PM delivers remarks on the 100th anniversary of the start of the First World War.* Ottawa, ON.

Hartung, W. D., & Pemberton, M. (2008). *Lessons from Iraq: Avoiding the next war.* Boulder, CO: Paradigm Publishers.

Harvey, David, *A Brief History of Neoliberalism* (Oxford, 2005; online edn, Oxford Academic, 12 Nov. 2020), https://doi.org/10.1093/oso/9780199283262.001.0001, accessed 5 Sept. 2025.

Hibou, B. (2004). *Privatizing the state.* New York, NY: Columbia University Press.

Holmes, O. (2019, December 12). Detectors, jammers and cyber-attackers: The rise of antidrone tech. *The Guardian*. https://www.theguardian.com/news/2019/dec/12/detectors-jammers-and-cyber-attackers-the-rise-of-anti-drone-tech

Hooper, C. (2001). *Manly states: Masculinities, international relations, and gender politics*. New York, NY: Columbia University Press. https://doi.org/10.7312/hoop12074

Horowitz, M. C. (2021). When speed kills: Lethal autonomous weapon systems, deterrence and stability. In T. S. Sechser, N. Narang, & C. Talmadge (Eds.), *Emerging technologies and international stability* (pp. 144–168). London: Routledge.

Hubert, H., & Mauss, M. (1981). *Sacrifice: Its nature and functions* (Rev. ed.). Chicago: University of Chicago Press. (Original work published 1964)

Hunt, K., & Rygiel, K. (2006). *(En)gendering the war on terror: War stories and camouflaged politics*. Burlington, VT: Ashgate. https://doi.org/10.4324/9781315564371

Hyndman, J. (2008). Whose bodies count? Feminist geopolitics and lessons. In C. T. Mohanty, R. L. Riley, & M. B. Pratt (Eds.), *Feminism and war: Confronting U.S. imperialism* (pp. 194–207). London: Zed Books.

Isin, E., & Rygiel, K. (2007). Abject spaces: Frontiers, zones, camps. In E. Dauphinée & C. Masters (Eds.), *The logics of biopower and the war on terror: Living, dying, surviving* (pp. 181–203). New York, NY: Palgrave Macmillan. https://doi.org/10.1007/978-1-137-04379-5

Isin, E., & Turner, B. (Eds.). (2002). *Handbook of citizenship studies*. London: Sage. https://doi.org/10.4135/9781848608276

Jay, N. (1992). *Throughout your generations forever: Sacrifice, religion and paternity*. Chicago, IL: University of Chicago Press.

Joachim, J., & Schneiker, A. (2012). Of 'true professionals' and 'ethical hero warriors:' A gender-discourse analysis of private military and security companies. *Security Dialogue, 43*(6), 495–512.

Johnson, C. (2008). American imperialism: Enabler of war. In M. Pemberton & W. D. Hartung (Eds.), *Lessons from Iraq: Avoiding the next war* (pp. 19–25). Boulder, CO: Paradigm Publishers.

Kaempf, S. (2018). *Saving soldiers or civilians? Casualty-aversion versus civilian protection in asymmetric conflicts*. Cambridge University Press. https://doi.org/10.1017/9781108551816

Kahn, P. (2008). *Sacred violence: Torture, terror and sovereignty*. Ann Arbor, MI: University of Michigan Press.

Kaldor, M. H. (1977) The significance of military technology. *Bulletin of Peace Proposals, 8*(2), 121–123. https://doi.org/10.1177/096701067700800204

Kaplan, C. (2017). DRONE-O-RAMA: Troubling the temporal and spatial logics of distance warfare. In C. Kaplan & L. Parkes, *Life in the age of drone warfare* (pp. 161–177). Durham, NC: Duke University Press. https://doi.org/10.2307.j.ctv121033r.10

Keenan, D. (2005). *The question of sacrifice*. Bloomington, IN: Indiana University Press.

Kinsey, C. (2006). *Corporate soldiers and international security: The rise of private military companies*. New York, NY: Routledge.

Kinsey, C. (2008). Private security companies and corporate social responsibility. In A. Alexandra, D. Baker, & M. Caparini (Eds.), *Private military and security companies: Ethics, policies and civil-military relations* (pp. 70–87). New York, NY: Routledge. https://doi.org/10.4324/9780203930830

Kivlan, T. (2008, April 30). *Witness cite prostitution, other contracting abuses in Iraq*. Government Executive. https://www.govexec.com/defense/2008/04/witnesses-cite-prostitution-other-contracting-abuses-in-iraq/26803/

Landler, M. (2016, May 14). For Obama, an unexpected legacy of two full terms at war. *The New York Times*. Retrieved from www.nytimes.com/2016/05/15/us/politics/obama-as-wartime-president-has-wrestled-with-protecting-nation-and-troops.html

Lasky, M. (2006). Iraqi women under siege. *Code Pink/Global Exchange*, 1–29.

Leander, A., & Van Munster, R. (2007). Private security contractors in the debate about darfur: Reflecting and reinforcing neoliberal governmentality. *International Relations, 21*(2), 201–216. https://doi.org/10.1177/0047117807077004

Lévi-Strauss, C. (1955). The structural study of myth. *The Journal of American Folklore, 68*(270), 428–444. https://doi.org/10.2307/536768

Leys, C. (2008). *Total capitalism: Market politics, market state*. Monmouth: Merlin Press.

Lieber, K. (2005). *War and the engineers: The primacy of politics over technology*. Ithaca, NY: Cornell University Press.

Linebaugh, H. (2013, December 29). I worked on the U.S. drone program. The public should know what really goes on. *The Guardian*. Retrieved from www.theguardian.com/commentisfree/2013/dec/29/drones-us-military

Lister, R. (2002). Sexual citizenship. In E. F. Isin & B. S. Turner (Eds.), *Handbook of citizenship studies*. London: Sage. https://doi.org/10.4135/9781848608276

Losey, S. (2016, October 24). Air Force offers bonuses up to $175,000 for drone pilots. *Air Force Times*. https://www.airforcetimes.com/news/your-air-force/2016/10/24/air-force-offers-bonuses-up-to-175000-for-drone-pilots/

Lucas, G. R., Jr. (2010). Postmodern war. *Journal of Military Ethics, 9*(4), 289–298. https://www.tandfonline.com/doi/full/10.1080/15027570.2010.536399?scroll=top&needAccess=true

Lutz, C. (2002). Making war at home in the United States: Militarization and the current crisis. *American Anthropologist, 104*(3), 723–773.

Mandel, R. (2004). *Security, strategy, and the question for bloodless war*. Boulder, CO: Lynn Rienner Publishers.

Manigart, P. (2006). Restructuring the armed forces. In G. Caforio (Ed.), *Handbook of the sociology of the military* (pp. 323–343). New York: Springer.

Mann, M. (1996). Nation-states in Europe and other continents: Diversifying, developing, not dying. In G. Balakrishnan (Ed.), *Mapping the nation* (pp. 295–316). New York: Verso.

Maogoto, J. N., Newell, V., & Sheehy, B. (2009). *Legal control of the private military corporation*. Basingstoke: Palgrave Macmillan. https://doi.org/10.1057/9780230583016

Martin, M. J., & Sasser, C. W. (2010). *Predator the remote-control air war over Iraq and Afghanistan: A pilot's story*. Minneapolis, MN: Zenith Press.

Masters, C. (2007). Body counts: The biopolitics of death. In E. Dauphinee & C. Masters (Eds.), *The logics of biopower and the war on terror: Living, dying, surviving* (pp. 43–57). New York, NY: Palgrave Macmillan. https://link.springer.com/book/10.1007/978-1-137-04379-5

Masters, C. (2008). Bodies of technology and the politics of the flesh. In J. Parpart & M. Zalewski (Eds.), *Rethinking the man question: Sex, gender and violence in international relations* (pp. 87–109). New York, NY: Zed Books. https://doi.org/10.5040/9781350222342

Mbembé, A. (2003). Necropolitics (L. Meintjes, Trans.). *Public Culture, 15*(1), 11–40.

McClintock, A. (1993). Family feuds: Gender, nationalism and the family. *Feminist Review, 44*, 61–80. https://doi.org/10.2307/1395196

McFate, S. (2016, August 12). America's addiction to mercenaries. *The Atlantic*. Retrieved from www.theatlantic.com/international/archive/2016/08/iraq-afghanistan-contractor-pentagon-obama/495731/

McMichael, P. (1990). Incorporating comparison within a world-historical perspective: An alternative comparative method. *American Sociological Review, 55*, 385–397. https://doi.org/10.2307/2095763

McMichael, P. (2009). Global citizenship and multiple sovereignties: Reconstituting modernity. In Y. Atasoy (Ed.), *Hegemonic transitions, the state and crisis in neoliberal capitalism* (pp. 23–43). London: Routledge. https://doi.org/10.4324/9780203884027

McSorley, K. (2013). *War and the body: Militarisation, practice and experience*. Routledge. https://doi.org/10.4324/9780203081419

Medina, D. (2014, June 22). Drone markets open in Russia, China and rogue states as America's wars wane. *The Guardian*. Retrieved from www.theguardian.com/business/2014/jun/22/drones-market-us-military-china-russia-rogue-state

Memmi, D. (2002). Public-private opposition and biopolitics: A response to Judit Sandor. *Social Research, 69*(1), 143–147. https://doi.org/10.1353/sor.2002.0014

Mencimer, S. (2011, July 7). Why Jamie Leigh Jones lost her KBR rape case. *Mother Jones*. Retrieved from www.motherjones.com/politics/2011/07/kbr-could-win-jamie-leigh-jones-rape-trial/

Mills, C. W. (1959). *The sociological imagination*. New York: Oxford University Press.

Mission: Readiness. (2010). *Too fat to fight [Position Statement]*. Retrieved from http://cdn.missionreadiness.org/MR_Too_Fat_to_Fight-1.pdf

Moskos, C. (2000). Toward a postmodern military: The United States as a paradigm. In C. Moskos, D. Segal, & J. A. Williams (Eds.), *The postmodern military*. Oxford: Oxford University Press.

Moskos, C., Williams, J. A., & Segal, D. (Eds.). (2000). *The postmodern military*. Oxford: Oxford University Press.

Motamedi, M. (2023, August 22). *Iran unveils attack drone capable of striking Israel*. AlJazeera. https://www.aljazeera.com/news/2023/8/22/iran-unveils-attack-drone-capable-of-striking-israel

Moyn, S. (2021). *Humane: How the United States abandoned peace and reinvented war*. New York: Farrar, Straus and Giroux. https://doi.org/10.1111/mepo.12754

Mutimer, D. (2005). Sovereign contradictions: Maher Arar and the indefinite future. In E. Dauphinée & C. Masters (Eds.), *The logics of biopower and the war on terror: Living, dying, surviving* (pp. 159–179). New York: Palgrave Macmillan. https://doi.org/10.1007/978-1-137-04379-5

Mutlu, C. E., & Salter, M. B. (Eds.). (2013). *Research methods in critical security studies*. New York, NY: Routledge. https://doi.org/10.4324/9781003108016

Nancy, J.-L. (1991). The unsacrificeable. *Yale French Studies, 79*, 20–38. https://doi.org/10.2307/2930245

Ong, A. (1999). *Flexible citizenship: The cultural logics of transnationality*. Durham, NC: Duke University Press. https://doi.org/10.2307/jj.13982283

Parenti, C. (2007). Planet America: The revolution in military affairs as fantasy and fetish. In A. Dawson & M. J. Schueller (Eds.), *Exceptional state: Contemporary U.S. culture and the new imperialism* (pp. 88–105). Durham, NC: Duke University Press. https://doi.org/10.1515/9780822389644

Parks, L., & Kaplan, C. (Eds.). (2017). *Life in the age of drone warfare*. Durham, NC: Duke University Press. https://doi.org/10.1515/9780822372813

Pattison, J. (2014a). *The morality of private war: The challenge of private military and security companies* (1st ed.). Oxford University Press. https://doi.org/10.1093/acprof:oso/9780199639700.001.0001

Pattinson, J. (2014b). The privatization of military force and the constraints on war. In *The morality of private war: The challenge of private military and security companies* (pp. 141–158). Oxford, UK: Oxford Academic. https://doi.org/10.1093/acprof:oso/9780199639700.003.0006

Pattison, J. (2024). Ukraine, Wagner, and Russia's convict-soldiers. *Ethics & International Affairs, 38*(1), 17–30. https://doi.org/10.1017/S0892679424000042

Patton Rogers, J. (2024). *De Gruyter handbook of drone warfare*. Berlin; Boston, MA: De Gruyter. https://doi.org/10.1515/9783110742039

Pearce, F. (2010). Obligatory sacrifice and imperial projects. In W. J. Chambliss, R. Mechalowski, & R. C. Kramer (Eds.), *State crime in a global age* (pp. 45–66). New York, NY: Routledge. https://doi.org/10.4324/9781843927051

Pemberton, M., & Hartung, W. D. (Eds.). (2008). *Lessons from Iraq: Avoiding the next war*. Boulder, CO: Paradigm Publishers. https://doi.org/10.4324/9781315633657

Pengelly, M. (2017, February 26). Father of Navy Seal killed in Yemen calls for investigation into 'stupid mission.' *The Guardian*. Retrieved from www.theguardian.com/us-news/2017/feb/26/father-navy-seal-yemen-trump-investigation-stupid-mission

Phillips, N., & Hardy, C. (2002). *Discourse analysis: Investigating processes of social construction*. Thousand Oaks, CA: Sage.

Pin Fat, V., & Stern, M. (2005). The scripting of Private Jessica Lynch: Biopolitics, gender, and the 'feminization' of the US military. *Alternatives: Global, Local, Political, 30*, 25–53. https://doi.org/10.1177/030437540503000102

Pugliese, J. (2013). *State violence and the execution of law: Biopolitical caesurae of torture, black sites, drones* (1st ed.). Routledge. https://doi.org/10.4324/9780203597743

Purkiss, J., & Serle, J. (2017, January 17). *Obama's covert drone war in numbers: Ten times more strikes than Bush*. The Bureau of investigative Journalism. https://www.thebureauinvestigates.com/stories/2017-01-17/obamas-covert-drone-war-in-numbers-ten-times-more-strikes-than-bush

Rabinow, P. (2006). Biopower today. *BioSocieties, 1*(2), 195–217. https://doi.org/10.1017/S1745855206040014

Rasheed, Z. (2023, January 24). *How China became the world's leading exporter of combat drones*. AlJazeera. https://www.aljazeera.com/news/2023/1/24/how-china-became-the-worlds-leading-exporter-of-combat-drones

Reagan, R. (1988, November 11). *Remarks at the veterans day ceremony at the tomb of the unknown soldier*. (Online by G. Peters & J. T. Woolley). The American Presidency Project. Retrieved from www.presidency.ucsb.edu/ws/?pid=35153

Rech, M., Bos, D., Jenkings, K. N., Williams, A., & Woodward, R. (2015). Geography, military geography, and critical military studies. *Critical Military Studies, 1*(1), 47–60. https://doi.org/10.1080/23337486.2014.963416

Reid, J. (2006). *The biopolitics of the war on terror: Life struggles, liberal modernity, and the defence of logistical societies*. Manchester: Manchester University Press. https://doi.org/10.7228/manchester/9780719074059.001.0001

Renic, N. C. (2020). *Asymmetric killing: Risk avoidance, just war, and the warrior ethos*. Oxford University Press. https://doi.org/10.1093/oso/9780198851462.001.0001

Reprieve. (2017, February 13). *America's deadly drones programme*. https://reprieve.org/us/2017/02/13/americas-deadly-drones-programme/

Richardson, M. (2022). How to witness a drone strike. *Digi War, 3*, 38–52. https://doi.org/10.1057/s42984-022-00048-3

Rose, N. (1999). *Powers of freedom: Reframing political thought*. New York, NY: Cambridge University Press. https://doi.org/10.1017/CBO9780511488856

Rosenberg, J. (1994). *The empire of civil society: A critique of the realist theory of international relations*. New York, NY: Verso.

Rupka, S., & Baggiarini, B. (2018). The (non) event of state terror: Drones and divine violence. *Critical Studies on Terrorism, 11*(2), 342–356. https://doi.org/10.1080/17539153.2018.1456735

Russell, S. (2020). *Human compatible*. London: Penguin Books.

Russon. (2016). https://www.ibtimes.co.uk/us-military-using-brain-zapping-technology-boost-mind-power-its-fighter-pilots-1590484

Rygiel, K. (2008a). The securitized citizen. In E. Isin (Ed.), *Recasting the social in citizenship* (pp. 210–239). Toronto, ON: University of Toronto Press. https://doi.org/10.3138/9781442688957

Rygiel, K. (2008b). *(En)gendering the war on terrorism: War stories and camouflaged politics* (K. Hunt, Ed.). London: Routledge. https://doi.org/10.1177/030981680809400113

Said, E. W. (1979). *Orientalism* (First Vintage books edition.). Vintage Books.

Salter, M., & Mutlu, C. E. (2012). *Research methods in critical security studies: An introduction*. New York, NY: Routledge. https://doi.org/10.4324/9780203107119

Sassen, S. (2002). Towards post-national and denationalized citizenship. In E. Isin & B. Turner (Eds.), *Handbook of citizenship studies* (pp. 277–291). London: Sage. https://doi.org/10.4135/9781848608276

Sassen, S. (2006). *Territory, authority, rights: From medieval to global assemblages*. Princeton, NJ: Princeton University Press. https://doi.org/10.1515/9781400828593

Sauer, F., & Schörnig, N. (2012). Killer drones: The 'silver bullet' of democratic warfare? *Security Dialogue*, *43*(3), 363–380. https://doi.org/10.1177/0967010612450207

Savage, C. (2022, October 7). White House tightens rules on counterterrorism drone strikes. *New York Times*. https://www.nytimes.com/2022/10/07/us/politics/drone-strikes-biden-trump.html

Scarry, E. (1987). *The body in pain: The making and unmaking of the world*. Oxford University Press.

Scharre, P. (2018). *Army of none: Autonomous weapons and the future of war*. New York: W. W. Norton.

Schott, R. (2010). *Birth, death, and femininity: Philosophies of embodiment*. Bloomington and Indianapolis, IN: Indiana University Press.

Schuck, P. (2002). Liberal citizenship. In E. Isin & B. Turner (Eds.), *Handbook of citizenship studies* (pp. 131–144). London: Sage. https://doi.org/10.4135/9781848608276

Schwarz, E. (2019). Günther Anders in Silicon Valley: Artificial intelligence and moral atrophy. *Thesis Eleven*, *153*(1), 94–112. https://doi.org/10.1177/0725513619863854

Shevchenko, V. (2023, October 21). *Ukraine fears drone shortages due to China restrictions*. BBC. https://www.bbc.com/news/world-europe-67078089

Shin, H. (2024, January 19). *North Korea says tests underwater nuclear drone, criticizes US-led joint drills*. Reuters. https://www.reuters.com/world/asia-pacific/north-korea-conducts-test-underwater-nuclear-weapons-system-kcna-2024-01-19/

Singer, B. C. J. (1996). Cultural versus contractual nations: Rethinking their opposition. *History and Theory, 35*(3), 309–337. https://doi.org/10.2307/2505452

Singer, B. C. J., & Weir, L. (2006). Politicd and sovereign power: Considerations on Foucault. *European Journal of Social Theory, 9*(4), 443–465. https://doi.org/10.1177/1368431006073013

Singer, B. C. J., & Weir, L. (2007). Politics and sovereign power: Considerations on Foucault. European. *Journal of Social Theory, 9*(4), 443–465. https://doi.org/10.1177/1368431006073013

Singer, B. C. J., & Weir, L. (2008). Sovereignty, governance and the political. *Thesis, 11*(94), 49–71. https://doi.org/10.1177/0725513608093276

Singer, P. (2004, November). *The private military industry and Iraq: What have we learned and where to next?* [Policy paper]. Geneva Centre for the Democratic Control of Armed Forces (DCAF). Retrieved from www.dcaf.ch/Publications/The-Private-Military-Industry-and-Iraq

Shaw, I. G. R. (2013, June 14). Predator empire: The geopolitics of US drone warfare. *Geopolitics, 18*(3), 536–559. https://doi.org/10.1080/14650045.2012.749241

Shimko, K. (2010). *The Iraq wars and America's military revolution*. Cambridge: Cambridge University Press.

Singer, P. (2005). Outsourcing war. *Foreign Affairs, 84*(2), 119–133.

Sluka, J. (2013). Virtual wars in the tribal zone: Air strikes, drones, civilian casualties, and losing hearts and minds in Afghanistan and Pakistan. In N. L. Whitehead & S. Finnström (Eds.), *Virtual war and magical death* (pp. 171–193). Durham, NC: Duke University Press. https://doi.org/10.2307/j.ctv125jjf1

Stravrianakis, A. (2016). Legitimising liberal militarism: Politics, law and war in the Arms Trade Treaty. *Third World Quarterly, 37*(5), 840–865. https://doi.org/10.1080/01436597.2015.1113867

Summers, C. (2004). Mandatory arbitration: Privatizing public rights, compelling the unwilling to arbitrate. *University of Pennsylvania Journal of Labor and Employment Law, 6*(3), 685–734. https://scholarship.law.upenn.edu/jbl/vol6/iss3/6/

Sylvester, C. (2012). War experiences/war practices/war theory. *Millennium Journal of International Studies, 40*(3), 483–503. https://doi.org/10.1177/0305829812442211

Tahir, M. (2017). The containment zone. *Life in the Age of Drone Warfare,* edited by LISA PARKS and CAREN KAPLAN, Duke University Press, pp. 220–40. JSTOR, https://doi.org/10.2307/j.ctv121033r.13. Accessed 11 Sept. 2025.

Taussig-Rubbo, M. (2009). Sacrifice and sovereignty. In J. Culbert & A. Saratker (Eds.), *States of violence: War, capital punishment, and letting die* (pp. 83–127). Cambridge, MA: Cambridge University Press.

Taussig-Rubbo, M. (2011). The unsacrificeable subject. In A. Sarat & K. Shoemaker (Eds.), *Who deserves to die?* (pp. 131–143). Boston, MA: University of Massachusetts Press.

Thompson, C. (2015, December 16). U.S. Air Force hires private companies to fly drones in war zones. *CorpWatch*. Retrieved from https://www.corpwatch.org/article/us-air-force-hires-private-companies-fly-drones-war-zones

Thompson, N., & Blanchfield, M. (2017, April 9). Canadians attend memorials for Vimy Ridge Centennial country-wide. *National Observer*. https://www.nationalobserver.com/2017/04/09/news/canadians-attend-memorials-vimy-ridge-centennial-country-wide

Troyer, L. (2003). Counterterrorism. *Critical Asian Studies, 35*(2), 259–276. https://doi.org/10.1080/1467271032000090143

Turnbull, T. (2024, June 4). *Australian army to allow recruits from foreign nations*. BBC. https://www.bbc.com/news/articles/cv22v0wg8v3o

Uesseler, R. (2008). *Servants of war: Private military corporations and the profit of conflict* (J. Chase, Trans.). New York, NY: Soft Skull Press. (Original work published 2006)

US Air Force. (n.d.). *Air Force distributed common ground system*. https://www.af.mil/About-Us/Fact-Sheets/Display/Article/104525/air-force-distributed-common-ground-system/

US Congressional Budget Office. (2008, August). *Contractors' support of operations in Iraq*. Washington, DC: Congressional Budget Office.

US Department of Defense. (2011). *Unmanned systems integrated roadmap: FY2011–2036*. https://irp.fas.org/program/collect/usroadmap2011.pdf

US Department of Defense. (2012, January). *Sustaining U.S. global leadership: Priorities for 21st century defense*. https://apps.dtic.mil/sti/pdfs/ADA554328.pdf

US Government. (2009, January). *U.S. government counterinsurgency guide*. https://2009-2017.state.gov/documents/organization/119629.pdf

US Government. (2014). *Unmanned systems integrated roadmap FY2013–2038*.

Vance, J. (1997). *Death so noble*. Vancouver, BC: UBC Press.

Virilio, P. (1989). *War and cinema: The logistics of perception*. Verso.

Wedel, J. R. (2008). The shadow army: Privatization. In M. Pemberton & W. D. Hartung (Eds.), *Lessons from Iraq: Avoiding the next war* (pp. 116–124). Boulder, CO: Paradigm Publishers.

Weisgerber, M. (2016, February 23). Back to Iraq: US military contractors return in droves. *Defense One*. Retrieved from www.defenseone.com/threats/2016/02/back-iraq-us-military-contractors-return-droves/126095/

Whitehead, N. L., & Finnström, S. (Eds.). (2013). *Virtual war and magical death: Technologies and imaginaries for terror and killing*. Durham, NC: Duke University Press. https://doi.org/10.2307/j.ctv125jjf1

The White House. (2015, December 14). *Remarks by the president on the military campaign to destroy ISIL* [Press release]. https://obamawhitehouse.archives.

gov/the-press-office/2015/12/14/remarks-president-military-campaign-destroy-isil

Whitworth, S. (2004). *Men, militarism and UN peacekeeping*. Boulder, CO: Lynne Rienner.

Wilcox, L. (2017). Embodying algorithmic war: Gender, race, and the posthuman in drone warfare. *Security Dialogue, 48*(1), 11–28. https://doi.org/10.1177/0967010616657947

Wilcox, L. B. (2015). *Bodies of violence: Theorizing embodied subjects in international relations*. Oxford University Press. https://doi.org/10.1093/acprof:oso/9780199384488.003.0002

Winter, J. M. (2006). *Remembering war: The Great War between memory and history in the twentieth century*. New Haven, CT: Yale University Press.

Wong, L. (2006). Combat motivation in today's soldiers: US army war college strategic studies institute. *Armed Forces and Society, 32*(4), 659–663. https://doi.org/10.1177/0095327X06287884

Woodward, R. (2008). Not for queen or country or any of that shit … Reflections on citizenship and military participation in contemporary British soldier narratives. In D. Cowen & E. Gilbert (Eds.), *War, citizenship, territory* (pp. 363–385). New York, NY: Routledge. https://doi.org/10.4324/9780203938126

Zehfuss, M. (2003). Forget September 11. *Third World Quarterly, 24*(3), 513–528. https://doi.org/10.1080/0143659032000084447

Zehfuss, M. (2018). *War and the politics of ethics* (1st ed.). Oxford University Press. https://doi.org/10.1093/oso/9780198807995.001.0001

Zenko, M. (2016, May 18). Mercenaries are the silent majority of Obama's military. *Foreign Policy*. Retrieved from http://foreignpolicy.com/2016/05/18/private-contractors-are-the-silent-majority-of-obamas-military-mercenaries-iraq-afghanistan/

Zenko, M. (2017, January 20). *Obama's final drone strike data*. Council on Foreign Relations. https://www.cfr.org/blog/obamas-final-drone-strike-data

Zuriek, E., & Hindle, K. (2004). Governance, security and technology: The case of biometrics. *Studies in Political Economy, 73*, 113–137. https://doi.org/10.1080/19187033.2004.11675154

Index